INFORMATION SECURITY HANDBOOK

Internet of Everything (IoE): Security and Privacy Paradigm

Series Editors: Vijender Kumar Solanki, Raghvendra Kumar, and Le Hoang Son

For more information about this series, please visit: https://www.routledge.com/Internet-of-Everything-IoE/book-series/CRCIOESPP

INFORMATION SECURITY HANDBOOK

Edited by
Noor Zaman Jhanjhi, Khalid Hussain,
Azween Bin Abdullah, Mamoona Humayun, and
João Manuel R.S. Tavares

CRC Press
Taylor & Francis Group
Boca Raton London New York

CRC Press is an imprint of the
Taylor & Francis Group, an **informa** business

First edition published 2022
by CRC Press
6000 Broken Sound Parkway NW, Suite 300, Boca Raton, FL 33487-2742

and by CRC Press
2 Park Square, Milton Park, Abingdon, Oxon, OX14 4RN

Library of Congress Cataloging-in-Publication Data
Names: Jhanjhi, Noor Zaman, 1972- editor. | Hussain, Khalid, editor. | Humayun, Mamoona, editor. | Abdullah, Azween Bin, 1961- editor. | Tavares, João Manuel R. S., editor.
Title: Information security handbook / edited by Noor Zaman Jhanjhi, Khalid Hussain, Azween Bin Abdullah, Mamoona Humayun, and João Manuel R.S. Tavares.
Description: First edition. | Boca Raton : CRC Press, [2022] | Series: Internet of everything (IoE): security and privacy paradigm | Includes bibliographical references and index. | Summary: "This handbook provides a comprehensive collection of knowledge for emerging multidisciplinary research areas such as cybersecurity, IoT, Blockchain, Machine Learning, Data Science, and AI. This book brings together in one resource Information security across multiple domains. Information Security Handbook addresses the knowledge for emerging multidisciplinary research. It explores basic and high-level concepts, serves as a manual for industry, while also helping beginners to understand both basic and advanced aspects in security-related issues. The handbook explores security and privacy issues through IoT ecosystem and implications to the real world and at the same time explains the concepts of IoT-related technologies, trends, and future directions. University graduates and postgraduates, as well as research scholars, developers, and end-users, will find this handbook very useful"-- Provided by publisher.
Identifiers: LCCN 2021043552 (print) | LCCN 2021043553 (ebook) | ISBN 9780367365721 (hbk) | ISBN 9781032203751 (pbk) | ISBN 9780367808228 (ebk)
Subjects: LCSH: Computer networks--Security measures.
Classification: LCC TK5105.59 .I527 2022 (print) | LCC TK5105.59 (ebook) | DDC 005.8--dc23/eng/20211110
LC record available at https://lccn.loc.gov/2021043552
LC ebook record available at https://lccn.loc.gov/2021043553

ISBN: 978-0-367-36572-1 (hbk)
ISBN: 978-1-032-20375-1 (pbk)
ISBN: 978-0-367-80822-8 (ebk)

DOI: 10.1201/9780367808228

Typeset in Times
by MPS Limited, Dehradun

Contents

Preface

We are living in a time of cutting-edge technology, where the world is receiving several types of challenges day by day. The current COVID-19 pandemic is one of the challenging scenarios, where the entire world is completely dependent on technology for keeping daily routines in action. Technology evolves in each domain of life, from health to education, government to private sectors, business to personal, etc. This huge level of technological involvement everywhere in life raised the demands of safety and security as well. Security gains higher importance now than ever before. Recently, a number of cybersecurity threats were examined globally for several online platforms that are helping to continue daily life during this pandemic.

This book is an attempt to collect and publish innovative ideas, emerging trends, implementation experience, and use cases pertaining to different enabling security approaches to keep the current technologies secure. Overall, this book aims to have a collection of ideas related to computer security, cybersecurity, network security, etc., and how this collection can contribute for the researchers and current literature to help to enhance security. The book is organized as follows.

The Organization of the Book

CHAPTER 1: SC-MCHMP: SCORE-BASED CLUSTER-LEVEL HYBRID MULTI-CHANNEL MAC PROTOCOL FOR WIRELESS SENSOR NETWORK

The potential applications of WSN (wireless sensor network) include smart spaces, environmental examininations, robotic study, and medical systems. To design the WSN, efficient energy is considered. Because of data transmission from sensor nodes, the collision occurs in WSNs, and the traffic is higher at SINK nodes owing to the excess data transmission at the sensor nodes. The MAC (medium access control) mechanism manages the important division of resources consumption. The single channel initiates for data transmission using WSNs in an existing MAC protocol. Also, the unexpected variation in the quality of the link in addition to the status of the node are caused by the deployment of WSN within the severe atmospheres. As a result of the changes in the status of the node, in addition to quality of the link, a change occurs in end-to-end delay of every sensor node. However, constrained energy is used to supply the sensor nodes, and it is used in extending the lifetime of the network. A new and simple routing mechanism, Scoreboard Cluster-Level Multi-Channel Hybrid MAC Protocol (SC-MCHMP) is proposed here for dealing with these issues; it utilizes the multi-channel MAC procedure comprising TDMA activity (time division multiple access) sequencing nodes and FDMA (frequency division multiple access) for collision-free exchange and CSMA/CA (Carrier Sense Multiple Access/Collision Avoidance) for data transmission. This method minimizes the energy consumption and reduces the nodes' overhead, achieving the collision-free transmission. The method to calculate the score and the score-based route selection increases the network lifetime and improves the network performance.

CHAPTER 2: SOFTWARE-DEFINED NETWORKING (SDN) SECURITY CONCERNS

Software-defined networking (SDN) is an approach of networking to enable network administrators to respond rapidly through a (logically) centralized controller. It may change the limitations of classical network infrastructure. SDN broke the vertical integration and separates the control plane (controller) from the data plane (switches) that forward the traffic. The main idea behind SDN is to decouple the data plane/forwarding plane from the control plane where the controller controls and manages the devices residing in the forwarding plane. The centrally controlled control plane benefits the SDN for being flexible and programmable with the control over the network traffic flows. The control and data plane are decoupled, and that could understood by way of a programming interface among SDN controller and switches. The network administrator can manage and control the network and packet-processing functions through his own program.

CHAPTER 3: CLUSTERING IN WIRELESS SENSOR NETWORKS USING ADAPTIVE NEURO-FUZZY INFERENCE LOGIC

Wireless sensor networks are powerful categories of mobile ad hoc networks that provide easy and efficient communication of technology and human. Wireless sensor networks are widely used for the results they provide, which include the reduction in human work. Sensors are deployed in groups in required areaa where nodes collect data from the surroundings and send it to the sink node through multi-node communication. This process involves lots of energy dissipation of nodes at individual levels, leading to the early fall of the network. To solve this problem, the concept of clustering was given in hierarchical routing protocols. The clustering process also lacks efficiency as cluster heads are selected randomly. Appropriate selection of cluster heads may prove to be an effective and logical way to regulate energy consumption and increase network life. This paper proposes an efficient neuro-fuzzy logic-based technique to improve energy consumption and network performance. The wise selection of cluster head will aid in data-transmission efficiency, increasing functioning to ensure network life in emergencies. Adaptive neuro-fuzzy logic helps in training the parameters to meet the requirements of becoming cluster heads. The candidate cluster heads parameters are tested against the training data, and the appropriate one is selected as head. The proposed technique is tested for different network cases and has shown good results in case of packet delivery ratio.

CHAPTER 4: SECURITY IN BIG DATA

Big data has gained popularity in recent decades, and it contributed to several application domains, where it changes the shape of existing business; each and every thing related to the analysis was brought in before us in different ways. For ages, individuals have used Google to ask questions, such as applications of big data in businesses to prosper, how it can assist an organization with succeeding, which technologies of big data are favored for this purpose, and several other related questions. A great deal has been said and written already about big data, however, the term itself stays unexplained. To be fair, we haven't established a prevalent definition of it, such as: Big data means it's big in this idea brings up another query of how big it is, how to measure it; is it in a terabyte, petabyte or even more? So, to resolve this ambiguity, a need to define big data arises. The hype of big data applications is based on its applications and providing solutions to different complex issues. However, the security remains always a major concern, especially in case of data. The organizations also are concerned for the security issues in the big data, as well. This chapter will elaborate the security concerns with big data.

CHAPTER 5: PREVENTION OF DOS/DDOS ATTACKS THROUGH EXPERT HONEY-MESH SECURITY INFRASTRUCTURE

Today, denial of service (DOS) and distributed denial of service (DDOS) attacks are rapidly increasing on the internet. DDOS attacks are used to overload the network infrastructure and services. Such kinds of attacks lead to the unavailability of services across networks. Honey pots can be used to ensure the continuous availability of services across networks. Honey pot is defined as a trap that mimics, notices, and records overall activities of the attacker and prevents attacks efficiently. In this way, malicious data will not route toward the production servers. The main purpose of this paper is to prevent DOS/DDOS attacks through expert honey-mesh security infrastructure. In this paper, software simulation tool DDOSSim is used to identify and simulate DDOS attacks via defense mechanisms.

CHAPTER 6: EFFICIENT FEATURE GROUPING FOR IDS USING CLUSTERING ALGORITHMS IN DETECTING KNOWN/UNKNOWN ATTACKS

In this paper, various feature grouping techniques are analyzed, using various machine-learning approaches to investigate their accuracy. Real-time traffic can be monitored for network attacks, which can be used in monitoring both the extrusion as well as the intrusion traffic. The main aim is to identify network attacks for providing future-proof software solutions, such that false alarms could be reduced and a more secure network could be made. The extrusion traffic detects attacks within the network and movement of data out from the network, whereas the intrusion detection system will monitor the incoming packets of data in the network, thus monitoring all the traffic inside as well as outside and providing a better solution to the existing system. The rules in the snort would also be optimized for better detection purposes. In this paper, an algorithm is proposed to enhance the chances to detect intrusion and will perform efficient and optimized data delivery in internal and external network. The proposed work will add a trust parameter to IDS by learning attack patterns in the future. This work can further be extended to application levels where decentralized nodes can be added to blockchain techniques to add trust among the newly connected and adjoining nodes.

Chapter 7: PDF Malware Classifiers – A Survey, Future Directions, and Recommended Methodology

Malicious software continues to pose a major threat to the cyber world. Text files are the most frequently used vectors to infect various systems using malware. In all this, to execute the attack, the intruder attempts to merge the malignant code with the benevolent text data. Due to its compatibility and lightweight characteristics, PDF (portable document format) is the most widely used file method of sharing documents. In today's world, attackers are using cutting-edge methods to obfuscate malware concealed inside document files. So, it is difficult for malware detection classifiers to effectively identify the text. To understand their design and working procedures, we surveyed different types of learning-based PDF malware classifiers. Also, we have described about the pdf document by which we can understand the working of malware. Finally, we conducted a recommended methodology on the basis of the literature survey and specified the future direction for the better classification results. This work is the extension of dissertation.

Chapter 8: Key Authentication Schemes for Medical Cyber Physical System

Cyber-physical system (CPS) opens further directions for innovative discussions to medical practitioners, researchers, and industrialists. It is an extensive field that integrates and provides in-depth collaboration of computation, communication, and control technologies. Medical CPS is a perceived term that evolved from health monitoring, which is obviously a sub-field of it. Medical CPS, like other sub-fields of CPS, uses sensing technologies for reliable data acquisition and extensively depends on wireless sensor networks (WSN) field for data transmission, storage, and control. This data is acquired through wearable or implantable medical devices, which are then used by a medical practitioner to diagnose and prescribe pertaining to health conditions of the patient. Therefore, it is considered as a substantial topic in CPS domain. From the last two decades, data security has emerged as the primary concern for researchers due to the open-access wireless network in CPS. However, several key authentication schemes have been proposed as possible solutions that strengthen the data-protection types against vulnerabilities in the existing system. But due to rapid development in CPS-related technologies, new challenges have been evolving that created further considerations to test multiple design variations for key authentication schemes. This chapter covers detailed analysis of the authentication schemes and compares their efficiency in terms of performance, storage, and computational requirements. Finally, it also highlights the pros and cons of each scheme of protection against available cyber-attacks in edical CPS.

Chapter 9: Ransomware Attack: Threats & Different Detection Technique

When we talk about cybersecurity, one of the threats that creates the major damage is ransomware. This attack is a one of many malware attacks; there are various others like worms and viruses, but ransomware is trending and one of the most dangerous threats over the network. Ransomware is not a new type of attack; it just evolved rapidly over time and as per the study with evolution, various detection techniques also evolved and machine learning, as well as deep learning, is the approach through which detection becomes better. Also, we see the payment method that plays an important role in the attack system; in this paper, we understand the analysis phase in the training of the system and also learn about the link or path from starting to the end of the attack; if someone wants to break the attack, then link or condition should break.

CHAPTER 10: SECURITY MANAGEMENT SYSTEM (SMS)

In the current era of information technology, most of the companies, industries, and government sectors have already automated their systems using computer networks. These networks primarily store 'data' of the organization for further processing and later use. In the information-security context, which is the ultimate insurance of the corporation's information, it is the core interest to protect and zealously preserve the information and, at the same time, the availability of the information when needed is a greater challenge. So, the network managers need a system to orientate facilities in place to cover data and network resources. This chapter explores the security management system (SMS), which is a comprehensive set of strategies and procedures based on the risk management and risk assessment components; it looks into all critical business processes by analyzing the associated risks. Afterward, it executes the controls to prevent information from internal and external attacks (threats) and ensures that information is well safeguarded, and risk is absolved. These threats are not only the cause of someone with malicious targets but also the accidental events like someone who downloads a Trojan or involuntarily deletes or moves critical files. The periodic check is the final step to adjust and organize the policies in the place they need to be, according to new technologies and new ways to save data damage externally. The essence of the implemented system (e.g. ISO/IEC 27001) based on the organizational management model Plan-Do-Check-Act (PDCA) framework, which embeds cloudy features to achieve the effectiveness and efficiency throughout the necessary procedures.

CHAPTER 11: AUTOMATIC STREET LIGHT CONTROL BASED ON PEDESTRIAN AND AUTOMOBILE DETECTION

Automatic street light control system is a system that is preferred over normal conventional street lights to save energy efficiently. Automatic street light control system is the system that makes use of the advanced automatic technologies to enlighten the road. The main consideration of this system is to find the amount of energy utilized and avoid wastage of electricity when a vehicle passes through the road because 30–40% of energy is being wasted by older street light systems at night. This system will start lighting with high intensity when vehicles or pedestrians pass through on the road, or on contrary, the lights will be in dim condition. Improvements in technology mean it is becoming eco-friendly in this advanced world. The advancement in an automatic technology in street lights is the optimum use of street lights and new techniques to produce much more efficient devices. This advancement in technologies results in overcoming the necessity of human resources. Automation plays a significant role within the world economy and in our day-to-day activities. Our Implementation work shows automatic street lights control systems give beneficial outcomes on which the power is utilized optimally and efficiently. The automatic street light with a microcontroller comes to the rescue in giving instructions to street lights when an automobile or pedestrian is detected; the IR signals are transmitted to the microcontroller, and the microcontroller performs necessary response to street lights. Therefore, we could efficiently utilize the power to light the street lamps. The lights will remain dim when there is no detection of automobiles or pedestrians. Automatic street light system uses 8051 Microcontroller AT89S52 to build this system, which performs based on the object detected in roads and performs necessary operations to streetlamps whether to switch light ON/OFF.

CHAPTER 12: COST-ORIENTED ELECTRONIC VOTING SYSTEM USING HASHING FUNCTION WITH DIGITAL PERSONA

Democracy is the anchor of the puissant and authoritative government. Fair, free, and transparent elections are crucial for electing competent leaders in democratic governments. But due to conventional election system rigging, conducting free elections is the most challenging task. Hence, system's security and authenticity of voter identity are the major problems. This chapter presents a proficient and secure electronic voting system. Here with we authenticate the licitness of the voter by voter ID and biometric fingerprints. The security of the system is obtained by VPN on specific domain. The adequate and effectual result is generated through the autocount process. The confirmations message is also sent for the satisfaction of the voter. The proposed system enhances the confidentiality, integrity, and authentication of the voting scheme; it also reduces the cost and improves the efficiency of the voting process, dynamically updating the results via graphical representation. This approach increases the voter's gratification and builds trust toward the election system.

CHAPTER 13: BLOCKCHAIN-BASED SUPPLY CHAIN SYSTEM USING INTELLIGENT CHATBOT WITH IoT-RFID

The blockchain is a peer-to-peer distributed ledger technology that provides trust and immutability. The data on blockchain is cryptographically secured, and it ensures the transparency, traceability, and quick updating. The artificial intelligence is the simulation of human process in an intelligent and smarter way. The chatbot helps us to provide the customer services and customer satisfaction. The IOT means different devices are connected together to exchange information. The RFID tag is used to track and verify the product. The RFID tags contains the digitally stored information about the product. The main problem is the traceability and how can we use these technologies to ensure the product originality and trace the product's journey. To ensure all these things, we are using all these technologies: the blockchain, artificial intelligence, and IoT to ensure the security, transparency, and traceability of the product in supply chain. We will be using RFID tags to verify the products and AI chatbot to provide the customer services. The data will be encrypted and secured on blockchain network. All this concludes that the blockchain with supply chain can give us good results. The system will be more efficient to ensure all the products and keeps that data confidential.

Editor Biographies

Noor Zaman Jhanjhi (NZ Jhanjhi) is currently working as Associate Professor, Director Center for Smart Society 5.0 [CSS5], and Cluster Head for Cybersecurity cluster, at the School of Computer Science and Engineering, Faculty of Innovation and Technology, Taylor's University, Malaysia. He is supervising a great number of postgraduate students mainly in the cybersecurity for Data Science. Cybersecurity research cluster has extensive research collaboration globally with several institutions and professionals. Dr Jhanjhi is Associate Editor and Editorial Assistant Board for several reputable journals, including IEEE Access Journal, PeerJ Computer Science, PC member for several IEEE conferences worldwide, and guest editor for the reputed indexed journals. Active reviewer for a series of Q1 journals, has been awarded globally as a top 1% reviewer by Publons (Web of Science). He has been awarded as outstanding Associate Editor by IEEE Access for the year 2020. He has high indexed publications in WoS/ISI/SCI/Scopus, and his collective research Impact factor is more than 300 points as of first half of 2021. He has international Patents on his account, edited/authored more than 29 research books published by world-class publishers. He has a great experience of supervising and co-supervising postgraduate students, an ample number of PhD and Master students graduated under his supervision. He is an external PhD/Master thesis examiner/evaluator for several universities globally. He has completed more than 22 international funded research grants successfully. He has served as Keynote speaker for several international conferences, presented several Webinars worldwide, chaired international conference sessions. His research areas include, Cybersecurity, IoT security, Wireless security, Data Science, Software Engineering, UAVs.

Khalid Hussain has completed his PhD. in Computer Science from Universiti Teknologi Malaysia, Malaysia. He has total 31 years of teaching, industrial and administrative experience nationally and internationally. He has an intensive background of industry as well as academic quality accreditation in higher education besides scientific research activities, he had worked as convener with higher education commission of Pakistan for academic accreditation (National Computing Education Accreditation Council) for more than a decade and earned NCEAC accreditation five time for two programs at computer science and information technology department The University of Lahore. He also worked for National Computing Education Accreditation Council (NCEAC) Pakistan for evaluating other universities computing programs. Up till now he has evaluated up to 31 universities. He has experienced in teaching advanced era technological courses including, Advance Networks, Cryptography and Network Security, Operating

System, Research Methods, Cloud Computing besides other undergraduate and postgraduate courses, graduation projects and thesis supervision.

Dr. Khalid Hussain is an active researcher in the domain of cyber security. He is reviewer of several well reputed indexed journals. He has authored several research papers in ISI indexed and impact factor research journals\IEEE international conferences three books and one book chapter. He has successfully completed more than **06** national and international funded applied research projects. He also served as Keynote speaker for several conferences around the globe. He also chaired international conference sessions and presented session talks internationally. He has strong analytical, problem solving, interpersonal and communication skills. His areas of interest include Cyber Security, Wireless Sensor Network (WSN), Internet of Things (IoT), Mobile Application Development, Ad hoc Networks, Cloud Computing, Big Data, Mobile Computing, and Software Engineering.

Azween Abdullah (Senior Member, IEEE) has been contributing to research, teaching, consulting and in administrative services to the institutions that he has been working for thus far. He is a professional development alumni of Stanford University and MIT and his work experience includes 30 years as an academic in institutions of higher learning and as director of research and academic affairs at two institutions of higher learning, vice-president for educational consultancy services, 15 years in commercial companies as Software Engineer, Systems Analyst and as a computer software developer and IT/MIS consultancy and training. Dr. Azween Abdullah has several patents under his name and been actively serving as expert reviewer and editorial board member for several high impact technical journals. Prior to joining Taylor's University, Dr. Azween Abdullah served as faculty member at Monash University Malaysia and Universiti Teknoloji Petronas. He also serves as the adjunct research professor at the Malaysian University of Science and Technology. He has guided twelve PhD students and ten Masters students under his supervision and have secured both international and local research funding of more than RM700k. He is a fellow of the British Computer Society and members of IEEE and ACM. Dr. Azween Abdullah's general research interests are in the areas of cyber security and trustworthy computing, formal models of computation, and bio-inspired and quantum computing. He has been consulting for some technology companies in content development and has done research on emerging areas of networked and quantum security. He has been involved with a number of government and semi-government organisations in Malaysia in the role of external consultant and currently works with several government linked companies and industries to promote cyber security capacity development. He has also published more than 150 publications in various technical journals and conference proceedings and has given technical talks in a number of key international conferences, industry summits and forums.

Mamoona Humayun, College of Computer and Information Sciences, Sakaka, Saudi Arabia. Mamoona Humayun received the PhD. degree in computer architecture from the Harbin Institute of Technology, China. She has 12 years of teaching and administrative experience internationally. She has supervised various master's and Ph.D. thesis. Her research interests include global software development, requirement engineering, knowledge management, cybersecurity, and wireless sensor networks. She is an active Reviewer for a series of journals.

João Manuel R. S. Tavares graduated in mechanical engineering from the Universidade do Porto, Portugal, in 1992. He received the M.Sc. and Ph.D. degrees in electrical and computer engineering from the Universidade do Porto, in 1995 and 2001, respectively, and the Habilitation degree in mechanical engineering, in 2015. He is currently a Senior Researcher with the Instituto de Ciência e Inovação em Engenharia Mecânica e Engenharia Industrial and an Associate Professor with the Department of Mechanical Engineering, Faculdade de Engenharia da Universidade do Porto. He is the co-editor of more than 40 books, and the co-author of more than 35 book chapters and 600 articles in international and national journals and conferences. He holds three international patents and two national patents. He has been a Committee Member for several international and national journals and conferences. He is the co-founder and the co-editor of the book series Lecture Notes in Computational Vision and Biomechanics (Springer). He has been a (Co-) Supervisor for several M.Sc. and Ph.D. theses and a Supervisor for several Postdoctoral projects. He has participated in many scientific projects as a Researcher and as a Scientific Coordinator. His main research interests include computational vision, medical imaging, computational mechanics, scientific visualization, human–computer interaction, and new product development. He is the Founder and the Editor-in-Chief of the Computer Methods in Biomechanics and Biomedical Engineering: Imaging & Visualization (Taylor & Francis) and the Co-Founder and the Co-Chair of the International Conference Series, including CompIMAGE, ECCOMAS Vip IMAGE, ICCEBS, and BioDental. More information can be found at www.fe.up.pt/tavares.

Contributors

Hasnat Ahmed
Department of Computer Science and
Information Technology
The University of Lahore Islamabad
Islamabad, Pakistan

Faraz Ahsan
Department of Computing and
Technology
Abasyn University
Islamabad, Pakistan

Nuzhat Akram
Barani Institute of Sciences Sahiwal
Campus
PMAS Arid Agriculture
University RWP
Punjab, Pakistan

Abdulellah A. Alaboudi
College of Computer Science
Shaqra University
Shaqra, Kingdom of Saudi Arabia

Amjad Ali
Department of CS
CUI
Lahore, Pakistan

Mustansar Ali Ghazanfar
University of East London
London, UK

Tariq Ali
CS Department
CUI
Sahiwal, Pakistan

Saud Altaf
Assistant Professor
University Institute of Information
Technology
PMAS-Arid University Rawalpindi
Rawalpindi, Pakistan

Divya Anand
School of Computer Science and
Engineering
Lovely Professional University
Phagwara, India

Prikshat Angra
School of Computer Science &
Engineering
Lovely Professional University
Phagwara, Punjab, India

Jyotir Moy Chatterjee
Assistant Professor (IT)
LBEF
Kathmandu, Nepal

Umar Draz
Department of CS
UoS
Sahiwal, Pakistan

Seema Gaba
Lecturer
Department of Computer
Science and Engineering
Chandigarh University
Mohali, India

J Gitanjali
School of Information Technology
Engineering
Vellore Institute of
Technology University
Vellore, India

Radhika Gupta
Lecturer
Department of Computer
Science and Engineering
Cognizant Bangluru, India

Mamoona Humayun
Assistant Professor
Department of Information Systems
College of Computer and Information
  Sciences
Jouf University
Sakaka, Kingdom of Saudi Arabia

Khalid Hussain
Barani Institute
Punjab, Pakistan

Saleem Iqbal
Assistant Professor
University Institute of Information
  Technology
PMAS-Arid University Rawalpindi
Rawalpindi, Pakistan

NZ Jhanjhi
Associate Professor and Director
CSS5 School of Computer Science and
  Engineering (SCE)
Taylor's University
Selangor, Malaysia

Kashif Naseer Qureshi
Senior Assistant Professor
Department of Computer Science
Bahria University Islamabad
Islamabad, Pakistan

Kavita
Department of Computer Science and
  Engineering
Chandigarh University
Mohali, India

Muazzam A. Khan
Department of CS
QAU
Islamabad, Pakistan

Shehneela Khan
CS Department
VU
Lahore, Pakistan

R. Pradeep Kumar
Student Electrical Electronic and
  Engineering
Vellore Institute of Technology
  University
Vellore, India

Sripada Manasa Lakshmi
Department of Computer Science and
  Engineering
Koneru Lakshmaiah Education
  Foundation
Vaddeswaram, Guntur
Andhra Pradesh, India

Mehwish Malik
School of Electrical Engineering and
  Computer Sciences
NUST
Islamabad, Pakistan

Muhammad Junaid Nazar
Lecturer
Department of Computer Science
COMSATS University Islamabad
Attock, Pakistan

Baibhav Pathy
Student
Electrical Electronic and Engineering
Vellore Institute of Technology
  University
Vellore, India

Sowjanya Ramisetty
Department of Computer Science and
  Engineering
KG Reddy College of Engineering and
  Technology
Hyderabad, India

Ravishanker
School of Computer Science &
  Engineering
Lovely Professional University
Phagwara, Punjab, India

Zia ur Rehman
University Institute of Information
    Technology (UIIT)
Pir Mehr Ali Shah Arid Agriculture
    University (PMAS-AAUR)
Rawalpindi, Pakistan

Muhammad Talha Saleem
Barani Institute of Sciences Sahiwal
    Campus
JV of PMAS Arid Agriculture
University RWP
Rawalpindi, Pakistan

Kashif Sattar
Assistant Professor
University Institute of Information
    Technology
PMAS-Arid University Rawalpindi
Rawalpindi, Pakistan

Rakhi Seth
National Institute of Technology
Raipur, India

Shahida
Arid Agriculture University Barani
    Institute of Sciences
Sahiwal, Pakistan

Khurram Shahzad
Department of Computer Science and
    Information Technology
The University of Lahore
Islamabad, Pakistan

Aakanksha Sharaff
National Institute of Technology
Raipur, India

Awadhesh Kumar Shukla
School of Computer Science and
    Engineering
Lovely Professional University
Phagwara, India

Monica Sood
Department of Computer Science &
    Engineering
Chandigarh University
Mohali, Punjab, India

R Sujatha
Associate Professor
School of nformation Technology
    Engineering
Vellore Institute of Technology
    University
Vellore, India

M N Talib
Papua New Guinea University of
    Technology
Lae, PNG

Imran Taj
Sr. Team Lead, Project Management
    Office
Information Systems Branch
Ministry of Attorney General and
    Public Safety Sector
BC Public Service
British Colombia, Canada

Noor ul-Ain
Barani Institute of Sciences Sahiwal
    Campus
JV of PMAS Arid Agriculture
    University RWP
Rawalpindi, Pakistan

Hina Umbrin
Barani Institute of Sciences Sahiwal
  Campus
JV of PMAS Arid Agriculture
  University RWP
Rawalpindi, Pakistan

Khalid Hussain Usmani
Professor
Department of Computer Science
Barani Institute of Sciences
Sahiwal, Pakistan

Sahil Verma
Department of Computer Science and
  Engineering
Chandigarh University
Mohali, India

N.S. Vishnu
School of Computer Science and
  Engineering
Lovely Professional University
Phagwara, India

Sobia Wassan
Department of Business School
Nanjing University
  Jiangsu, China

Sana Yasin
Department of CS
UO Okara
Punjab, Pakistan

1 SC-MCHMP: Score-Based Cluster Level Hybrid Multi-Channel MAC Protocol for Wireless Sensor Network

Sowjanya Ramisetty, Divya Anand, Kavita, Sahil Verma, and Abdulellah A. Alaboudi

CONTENTS

1.1 INTRODUCTION

Energy efficiency plays a significant role in the conventional WSNs for improving the network lifetime with the requirements of quality of services (QoS), such as delay and bandwidth conditions, which are less significant. Nevertheless, reliability and the ability to realistically detect events are required by the existing WSN multimedia implementations, such as vehicular flow on the highways for the intelligence of the battlefield. A large amount of information is produced

DOI: 10.1201/9780367808228-1

within a shorter duration, which leads to a higher channel-contention degree and a higher possibility of the packet collision. However, WSNs are designed for various MAC Protocols [1] for improving energy efficiency, scalability, and throughput. Various healthcare implementations of WSN produce bursty traffic, generating lower delay, higher throughput, and higher delivery rate [2]. WSNs cannot provide the trustworthy and on-time data delivery using a single channel with higher data-rate requirements due to the collisions and constrained bandwidth. Therefore, the investigators from the past few years who support the bursty traffic within WSN were attracted by the multichannel communication.

An increase in the node density (no. of nodes/unit area) takes place when the implementation of WSN is comprehensive. This approach offers a novel challenge in designing a MAClayer protocol [3]. A single frequency (channel) is used, and the individual performance is limited by the application of the single-channel limits using various conventional MAC protocols. Moreover, WSN performance is increased by proposing various multichannel MAC-layer protocols. Compared to the single-channel MAC protocol within the entire set of traffic conditions, the multichannel MAC protocols were found to be [4] less energy efficient. The effective substantial outcome is not provided by the single-channel protocol due to the interference that occurs whenever an increase in the density of the network occurs. Designing the effective multichannel MAC protocol for achieving higher performance, as well as energy efficiency within various traffic conditions, is significant. The impacts of the interference, as well as the assertion upon the wireless medium through the scheduling-interfering transmission across various frequency channels for improving the throughput, is mitigated by an effective approach, i.e., multichannel communication.

WSN faces a significant challenge in energy consumption [5], which impacts the lifespan of WSN. Therefore, the data collection, data aggregation, and data transmission require additional energy. The energy gets depleted, although the sensor node is within a sleep condition when it is not processing. Hence, the low-powered balanced clusters that consider the end-to-end delay, as well as the cost functions, are constructed by demanding the efficient energy-aware routing methods. The data received out of the cluster members are aggregated by every Cluster Head (CH) within the inter-cluster communication [6], and the data is forwarded toward the sink node through the relay nodes. The intracluster communication is performed in the cluster wherein the data is gathered by every single node and sent toward the respective CH. The aggregated data that is sensed is transmitted toward the sink by the routing process, where a predetermined path is found.

WSN faces a significant challenge named as energy consumption [5] that impacts the lifespan of WSN. Therefore the data collection, data aggregation as well as data transmission requires additional energy. The energy gets depleted although the sensor node is within a sleep condition or when it is not processing. Hence, the low powered balanced clusters that consider the end-to-end delay, as well as the cost functions, are constructed by demanding the efficient energy-aware routing methods. The data received out of the cluster members are aggregated by every Cluster Head (CH) within the inter-cluster communication [6] and the data is forwarded towards the sink node through the relay nodes. The intracluster communication is performed in the cluster wherein the data is gathered by every single node and sent towards the respective CH. The aggregated data that is sensed is transmitted towards the sink by the routing process where a predetermined path is found.

The data combination associated with a particular situation is involved by the data aggregation within WSNs. The main aim of aggregation is to extend the lifetime of the network by reducing the consumption of resources, such as battery power or bandwidth, in addition to the transmission packets. Moreover, the quality of the service requirements, including data accuracy, latency, fault tolerance, and security [7], is affected by the data-aggregation process. Energy-efficient procedures around the data aggregation are required because of the wide-ranging applications of WSNs. Hence, a significant role is played by the application of the energy-aware programs, as well as algorithms [8,9].

The data aggregation is achieved by a well-known approach where the classification of the sensor nodes is done within various clusters and a CH is chosen within every single cluster to aggregate the tasks. The distribution of load among the nodes is ensured by CH by turning around the nodes. Recently, the clustering protocols have gained a lot of attraction from several investigators. The scalability and balancing of the loads are increased and the WSN lifetime is extended by an efficient method called clustering [10,11]. An assumption is made by such methods that the respective data is sent by the node toward the BS; nevertheless, it is an energy-inefficient approach because of the long-distance transmission of data. Certain nodes are allowed to be dependent upon the respective data in various hops toward the BS for solving such issues [12]. Certain power wastages exist within this stage since few control packets for the route construction are retransmitted by certain control packets.

Wireless sensor nodes are a significant factor of smart infrastructure, having a simpler, less costly structure and being appropriate to wide-ranging deployment situations. Therefore, it completes the smart-grid monitoring implementation [13–15]. Smart grids, environmental monitoring, surveillance operations, both home and industrial automation [16–18], and social networks, including mobile social networks [19–21], are the various applications of WSNs. Smaller sensors with constrained resources for processing power, data storage, and radio transmission [22,23] are included in the network. Constrained energy is a significant limitation in WSNs [24–26]. The energy is replenished impractically, as well as unfeasibly, on behalf of various implementations. Designing an energy-effective application system having a longer lifespan is complicated [26,27]. The substantial study was performed upon the routing-algorithm optimization in WSNs [28], data fusion [29], MAC optimization [30], and cross-layer optimization techniques where various levels are combined [31]. MAC optimization is classified among the substantial approaches. Using the sensor nodes for switching the working units off cyclically [31–35] is performed by the famous MAC optimization approaches. The consumption of energy of the operating units during the switched off (sleep) mode is less with one order of magnitude than the working (active) state. The nodes must be kept in a sleep condition for a longer duration for saving energy.

The proposed SC-MCHMP protocol selects the appropriate forwarder nodes by enhancing the forwarder node-selection method of the routing protocol by introducing multiple valid parameters such as node ID, the residual energy level, the distance, the delay, RSSI (Received Signal Strength Indicator), RDR (remaining delivery ratio), and the ETX (Expected Transmission Count) of the nodes. Based on the aforementioned parameters, the protocol estimate and assign a score for every sensor node. The score is a factor describing the stability of the sensor node. The high score of the node denotes the node's high stability. The residual energy, distance between the neighbours, node delay, and RSSI are the primary parameters that can be extracted directly from the nodes by exchanging the control packets of the routing protocol. The parameters such as RDR and ETX are estimated during run time and evaluated based on the node data transmission. These estimated values are then exchanged and initially compared with the 1-hop neighbour nodes, and the routing table of the nodes is updated accordingly. During data transmission, the nodes exchange the table information and calculate the current values of the neighbours considered for score calculation.

- We proposed an optimization protocol for WSNs to enhance the QoS and improve network energy optimization.
- The proposed protocol achieves a high data-delivery rate with reduced network overhead.
- The network end-to-end delay reduces up to 60% using the proposed protocol.
- The proposed protocol maintains the energy-dissipation level of the sensor nodes at a very low level.
- This protocol improves the forwarder node-selection process by selecting the appropriate neighbors using multiple valid parameters.
- The use of hybrid-MAC protocols ensures collision-free data transmission, even in burst traffic.

1.2 LITERATURE SURVEY

Various clustering methods were introduced for distributing the load among the nodes, as well as coping with the energy-constraint issue. A significant, classical distributed procedure within this field is LEACH [36]. The setup phase and the steady-state phase are the two phases included in this approach. Every single node tries to be a CH having a probability within the setup phase. The sensed data is transmitted by every single cluster member toward the corresponding CH within the steady-state phase. Data is transmitted by every single CH toward the BS later in the data aggregation. The chances for playing the CH role are possessed by the nodes having less battery power; this is because the chance of being a CH of every single node is possessed independently depending upon the probability value. Therefore, it doesn't ensure a suitable distribution of CHs.

The application of LEACH is done as a standard for proposing the enhancing approaches like LEACH-C [37], centralized clustering algorithms, and LEACH-E [38] and LEACH-B [39] distributed-clustering procedures that concentrate on the energy-consumption reduction with the help of the node's residual energy and various relevant standards. The organization of the network is done in even dimensions by the authors in [40]. A CH, next head and a set of sensor nodes are involved in every single cluster. A CH is selected depending upon the received data out of the sensor nodes, and the network is distributed by the BS within clusters. The creation of the set of the qualified succeeding heads having residual energy above the threshold value is done for taking the task of CH in the succeeding level. Additional power maintenance within the sensor node is possessed by this approach.

There are two fog-enabled VANET schemes; one is SIVNFC (Secure intelligent vehicular network using fog computing) and the other is SOLVE (localization system frameworks) [41]. The authors provide an overview of FANET, with its routing techniques, routing protocols and cloud computing [42] applications. A number of challenges and solutions are also addressed.

The authors present an energy-effective clustering approach for the hetero-geneous WSNs (EEHC), depending upon the weighted election probability of every single node to be a CH in [43]. Electing the appropriate CHs distributive alongside heterogeneity hierarchical WSNs is the objective. Ultimately, it achieves reasonable outcomes [44].

Article [45] proposes a distributed randomized clustering technique to generate a CH hierarchy. The node's energy increases by increasing the number of levels. Therefore, it causes a longer delay in the algorithm.

Selecting the smaller set of the nodes, considering the tasks of CH where the entire network is covered, is the main objective of DEECIC [46]. An exclusive ID assignment technique, dependent upon the information of every single local node, is developed by it. It keeps the network coverage during the network lifetime extension using 2-hop intra-cluster communication in addition to a good load distribution among the nodes.

In [47], the proposed model has analyzed the dynamic enforcement of permis-sions to a specific application based on the defined context, without the intervention of users. According to the functional groups, profiles have assigned to different applications using this model, and a set of permissions with some associated context is contained in these profiles.

In [48], the time and spatial-based traffic data is used and is extracted based on LSTM and CNN networks for improving the model accuracy. The near-term traffic details, including the speed, are detected using the attention-based model, which is essential to predict the flow future value.

Guo et al. [49] implements a clustering method using a medium access control (MAC) approach using a random contention model. An extended delay-based tech-nique for reducing the consumption of energy, as well as the response period within routing, is presented by Enokido et al. [50]. A data-aggregation approach for de-creasing the communication collision probability, as well as for increasing the energy-saving effects of the nodes, is presented by Hsieh et al. [51]. Service providers could perhaps flexibly and easily deliver their software and services to meet unique demands of a variety of services, including virtual and augmented reality, video gaming, e-health, and several others [52]. The amalgamative sharp WSN algorithm [53] is introduced to show the performance based on information transfer, routing, processing time, and energy calculation. For better results, the ML algorithms are adopted with the proposed algorithm. Results show the performance of the ASWSN algorithm.

An adaptive fuzzy-clustering technique wherein the fuzzy c-means protocol is implemented for balancing the clusters, wherein the cluster-based routing is performed, is described by Mohammad et al. [54]. In [55], a secure and lightweight IoT-based framework is contributed as a technology based on WSNs. The proposed security approach based on a COOJA simulator is compared with the existing security solutions like SIMON and SPECH.

In [56], the hybrid logical-security framework (HLSF) is proposed to provide the data confidentiality and authentication in IoT. A lightweight cryptographic mechanism is used by HLSF for unique authentication. The security level is improved, and the better functionalities of a network are provided by implementing energy-efficient schemes.

In [57], a comprehensive review of the machine-learning techniques application are presented for big data analysis in healthcare. The existing techniques advantages and drawbacks are highlighted, in addition to the different research challenges.

An optimal routing approach on behalf of WSN using optimum energy consumption is discussed by Huang et al. [58]. Low Energy Adaptive Clustering Hierarchy (LEACH) protocol [59] is the primary model wherein several benefits are derived among the entire existing techniques; these include energy efficiency, load balancing among the clusters, and the enhanced throughput by simple design.

This paper proposes an energy-efficient hierarchical cluster-based routing technique on behalf of WSN distributive. The selection of CHs is done at the edges of the trees, depending on the efficient local data, which is the major challenge in this method while constructing a routing tree. The CH decision-making technique uses various criteria, including the residual energy of every single node and the distance toward the BS alongside the developed routing tree. The decision of becoming a CH is decided by every single node. The suitable CHs are joined by the common nodes. A reduction in the control packets is done, which in turn saves the power within every single sensor node [60]; this is done by combining the routing and clustering essential approach.

The basic process of [61] E-voting, IOT with blockchain financing, and blockchain technology helps for providing the highest performance in rural areas. The [62] characteristics and techniques of some chaotic maps used to encrypt images were reviewed. Also, for images like boat, airplane, peppers, lake, and house, chaotic encryption is applied and analyzed; a comparative study of the different protocol evaluates in increasing the throughput and network efficiency [63].

By implementing efficient routing within WSNs, this paper proposes a novel Delay Constrained Energy-efficient Multi-hop Routing Algorithm (DCEMRA). A novel method termed as the delay-constrained reliable routing technique, where the energy consumption is reduced with the construction of the effective clusters exclusive to the increase in the end-to-end delay, is introduced in this approach. The introduction of the novel computing approaches is done in this model to find the distance amid the cluster members, as well as CHs, a single CH toward another CH, and the CHs toward the sink node. Furthermore, this method obtains the optimal shortest path by investigating the maximum-paths having a shorter distance, and the best shortest path with the help of a reducing factor is found by applying the rules. Additionally, the best, as well as the most suitable, paths out of the overall shorter paths, considering energy, distance, and delay parameters, are found by following the rules. Including the enhancement of the reliability, packet delivery ratio, network lifetime, and the reduction within energy consumption & delay [64] is the main benefit in this routing protocol.

The concept of Artificial Neural Network (ANN) is used as a deep learning technique for protection against dual attacks for GHA and BHA, in addition to the method of swarm-based Artificial Bee Colony (ABC) optimization [65].

The underlying protocol requires the fulfillment of the stringent requirements for providing seamless and interoperable communication in IoMT [66]. It's a challenging task that the multimedia sensors' heterogeneous nature makes interoperable. A comprehensive review of the existing protocol stacks of IoMT is provided for understanding the challenges faced by interoperable and seamless communication in IoMT. Their feasibility is analyzed for multimedia streaming applications.

1.3   PROPOSED WORK

The challenges in the high data rates, in addition to burst traffic within the multi-hop WSNs, are addressed here; huge packets must be transmitted toward the final destination within the hostile atmospheres. However, constrained energy is used to supply the sensor nodes, and it is used to extend the lifetime of the network. To address all these problems, we introduce a SCORE BASED CLUSTER LEVEL MULTI-CHANNEL HYBRID MACPROTOCOL (SC-MCHMP), which utilizes the multi-channel MAC procedure comprised of FDMA and TDMA for collision-free exchange and CSMA/CA for data exchanging; this enhances the throughput of the network whenever the overall traffic is sent toward sink. The proposed protocol uses node ID, the residual energy level, and the distance, the delay, RSSI, RDR (remaining delivery ratio), and the ETX value to filter out the best nodes based on the score of the link and the route, which is calculated from the all the above said parameters. The channel-allocation model for keeping the sink interfaces within the reception mode, in addition to a single neighbor of every interface within the transmission mode, is adjusted to optimize the sink reception. Therefore, the sink for receiving the data frames is constantly allowed by the channel-allocation approach. SINK is the initial node that must be activated, which is upon depth zero within the Network creation, beacon propagation, and neighbor discovery phase. A beacon that is propagated within a multi-hop method for reaching the overall network nodes is broadcast using a SINK. Bitmaps for representing the neighbors are used to avoid the network overloading using the long control messages for exchanging neighborhood data. A bitmap, which represents the overall nodes within the network, is constructed by each node with the help of a local propagation order. The node address having the similar index within the propagation order corresponds with every single index of the bitmap. After receiving a beacon by a node out of the additional node, a consideration is made that it is a neighbor. The bitmap of the respective 1-hop neighbors within the beacon is included in every single node for constructing 2-hop neighbor lists. After receiving the overall beacons out of the respective neighbors, the list of corresponding 2-hop neighbors are constructed. The bitmap of the respective 1-hop, as well as 2-hop neighbors within the beacon, is included in every single node for constructing 3-hop neighbor lists. The exchange of neighborhood data is done in an effective manner, excluding the collision, including the light overhead with the help of bitmap codification, and transmitting beacons within a TDMA model. The medium is accessed and the data is sent by CSMA/CA technique when the overall nodes are ready for communicating by the similar node competing upon the channel. Once the node having the smallest address among the corresponding 3-hop-neighborhood, which is not allocated with a channel, is ready, the node proceeds toward the respective channel allocation. The nodes that finished the channel-allocation technique for knowing the chance of choosing a channel are announced by using a bitmap, which is used to represent the overall nodes. After the selection of a channel by a node, it is broadcast within the beacon frame.

1.3.1   PROCEDURE OF ROUTING MECHANISM

The routing mechanism consists of four phases:-

- 1-hop discovery
- Metric calculation
- Link score evaluation
- Source selection

1.3.1.1   1-Hop Discovery

A HELLO message that contains the individual node ID is advertised by every single sensor node toward the additional nodes within this 1-hop discovery phase. The data of the respective neighbors is stored by every single sensor node, and a neighbor list is built after the reception of "Hello" messages. The ETX amid every sensor node pair is calculated by using a "Hello" message, which is of smaller size.

1.3.1.2   Metric Calculation

The data out of the neighboring nodes is required by the metric calculation. A message for the corresponding nodes is sent by every single node within the metric-calculation phase. The node ID, the residual energy level, and the distance, the delay, RSSI, RD (remaining deliveries), and ETX value are included in this message. The inspection of every message is done, and the storage of the data is done within a neighboring table. The routing parameter for every single neighbor is calculated by every single node, and the parameters in the neighbor table are stored. After receiving the novel messages out of the neighboring nodes, the neighbor table is updated continuously.

1.3.1.3   Link Score Calculation

The link score and the route score are the two scores included where the individual link amid two neighboring nodes refers to the link score, and the addition of the link scores within a routing path is referred to the route score. The application of the link score can be done as RD (remaining deliveries) value and the reverse of RD value. Equations (1.1) and (1.2) given below presents the link score Clk and route score Crt.

$$S_{link}(i, j) = \tfrac{1}{RD_{ij}} \tag{1.1}$$

$$S_{route}(i) = min\{S_{route}(j) + S_{link}(i, j)\} \tag{1.2}$$

Where $S_{link}(i, j)$ the score of the link

$S_{route}(I)$ represents the $node^n$ route score within the respective routing path. The minimal values as the route score are obtained using Equation (1.2). A minimum route score is presented for a sink node, which is fixed as zero. A reduced number of iterations is required for calculating the route score. The route score for the corresponding neighbor nodes that are serving as next-hop candidates are calculated after the calculation of the parameter and the optimal score are determined. The optimal score is advertised by every single sensor node upon updating the score.

1.3.1.4 Source Selection

The score of the scores for each of the corresponding neighbor is updated by every node, and the optimal score is found. The updating of the scores is done perfectly after various iterations, and the selection of the neighbor with the optimal route score is done as the parent node for forwarding the data sensed. The overall sensor nodes out of the respective route scores are updated to determine the routing paths.

1.3.2 CSMA/CA

A distributed coordination function (DCF) to share the access toward the medium, depending upon the CSMA/CA protocol, is defined by the IEEE 802.11 standard for WLAN. Because the channel, as well as the transmission data, is not detected by the node at the same time, the application of collision detection is not done. The node for the transmission is determined when a node listens to the channel. Two methods are used to perform the carrier sensing, wherein the activity upon the radio inter-ference is detected by physical carrier sensing and the DCF RTS/CTS access mode is used to perform virtual carrier sensing.

1.3.3 TDMA

The distribution of the bandwidth is done among several stations, depending upon the time within TDMA. While sending the data, the allocation of time is done by every single station, and the respective data is transmitted by every station within the time slot allocated. The commencement of the respective slot, in addition to the slot location, should be known by every single station. The synchronization among various stations is required by TDMA.

1.3.4 FDMA

Several frequency bands are obtained by dividing the available bandwidth within FDMA. The data is sent by allocating a band by every station where the reservation of the band is done for the entire duration. Smaller bands of the remaining fre-quency are used to separate the frequency bands of various stations. The loads out of various low-bandwidth channels are combined by FDMA and are transmitted with the help of a channel with high bandwidth.

Algorithm for Score calculation & Route selection

---

d_{ij}= distance; E_{res} = residual energy

ETX = expected transmission count; D_{ij} = delay of the link

$S_{link}(i,j)$= link score of the link(i,j) ; $S_{route}(i)$ = route score of route (i)

FSR = final selected route For all the nodes 'n'

        1-hop discovery phase

          Broadcast HELLO

          packets

            Update neighbour node table

  End

  For each 1-hop neighbour node n

          Calculate distance d_{ij}

$$d_{ij}=\sqrt{(x_j-x_i)^2+(y_j-y_i)^2}$$

          Estimate E_{res}

$$E_{res}= (E_{initial}- E_{consumed})$$

          Estimate ETX

          Estimate D_{ij}

$$\text{Estimate } RD= \frac{E_{res}}{ETX_{ij} \times E_{tx}(l, d_{ij}) \times D_{ij}}$$

  End for

  For each node n and link l

      Estimate link score $S_{link}(i,j)=RD_{ij}\frac{1}{}$,

        Estimate $S_{route}(i,j)$

        If ($S_{route}(i)>S_{route}(j)$)

            FSR = (i)

      Else

      End If   FSR = (j)

  End for

  For each FSR

      Assign nodes in CSMA

        mode If (data = true)

          Member nodes transmit the data using TDMA

        End If

        If (collision = true)

          Switch to FDMA

        End If

        End For

1.4 RESULTS AND DISCUSSION

Network Simulator-2 is used to implement this method using the simulation patch files. The experimental verification is performed by using NS-2. An open-source option, as well as working at packet level, is provided by an open network simulation program. The users have executable commands that use the input argument named as Tcl scripting language. The name of the Tcl script is fed as the input argument of NS-2 executable command of NS by the users. The creation of the simulation trace file is most of the circumstances.

The outputs of NS-2 might be text-based or animation-based afterward on the simulation. Tools like nam and xgraphs are used for interpreting such outcomes in a graphical or interactive way. An appropriate subset of text-based information is extracted by the users for analyzing specific network performance and is transformed toward a possible presentation.

Clustering and routing are performed by using 21 sensor nodes within this simulation. This method outperforms the previous methods, as shown in the simulation.

The system parameters utilized in the simulations are shown in Table 1.1. This method uses application traffic as CBR (Constant Bit Rate), and it supports the traffic control in the network. The routing protocol serves as AOMDV, and the application of it is done for routing within the network. DEEH-CH, DCEMRA, and SC-MCHMP are the routing techniques that are implemented to perform outcomes of the network. With the consideration of packet size as 1024 bytes, the transmission rate is 1024 bytes/0.1 ms, having a maximum speed of 35 m/s in addition to the total simulation time of 20 sec.

The end-to-end delay versus simulation time is illustrated in Figure 1.1, which includes comparing proposed protocol SC-MCHMP and previous techniques like DCEMRA and DEEH-CB. The SC-MCHMP outperforms in terms of reduced end-to-end delay for communication among nodes. The multichannel allocation

TABLE 1.1
Simulation Table of Proposed Network

Network Parameter	Value
Application Traffic Protocol	CBR
Rate of Transmission	1024 bytes/0.1 ms
Radio Range	250 m
Data Packet Size	1024 bytes
Channel data rate	40 Mbps
Maximum speed	35 ms/s
MAC protocol	MAC/802_11, MAC/CSMA/CA, Mac/TDMA, MAC/FDMA
Routing protocol	AOMDA
Network Simulation time	20 secs
Number of nodes used in simulation	21
Network Area	500 × 500
Routing methods	DEEH-CH, DCEMRA, SC-MCHMP

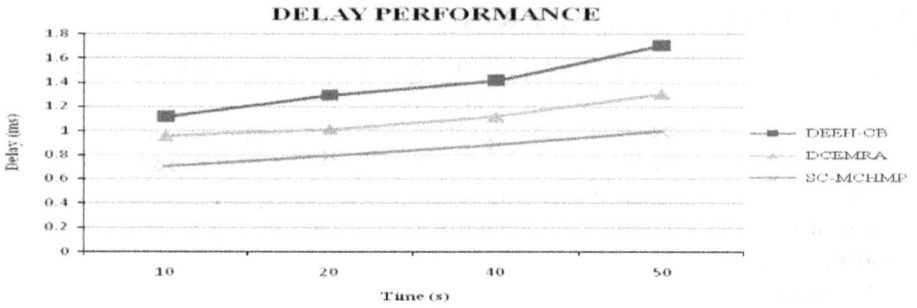

FIGURE 1.1   Performance on delay.

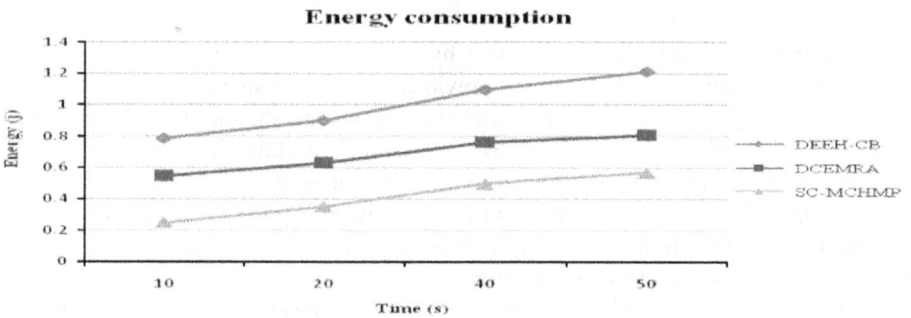

FIGURE 1.2   Energy consumption.

accomplishes using the CSMA or CA, TDMA modes, and this method has a 25% delay compared to the previous approaches. The maximum score of the route (in SC-MCHMP, less delay yields higher scores) ensures fewer delay in the selected paths, and more number of packets are getting delivered within a short time.

The simulation results of energy consumption are shown in Figure 1.2, which displays the graphs of energy level ratio vs simulation time. Based on the proposed protocol, the loss ratio improves with the reduction of energy conservation compared to the existing methods, such as DCEMRA and DEEH-CB. The collision-free data transmission out of the CH nodes toward the SINK results in reducing energy consumption and energy levels by 28%. Due to the reduced delay and the reduced number of retransmissions, the energy consumes less than the existing protocols.

The network performance is discussed in Figure 1.3, which presents the graphs of simulation time versus throughput. In this approach, a network's performance improves where the data saves more than the previous methods. This paper includes the throughput with an average of 34 Kb/s. Compared with the earlier approaches, the data transmission is done exclusively to the loss, and the network performs effectively. The SC-MCHMP always suggests less delay and interference-free paths. The packet delivery to the target node is mostly not disturbed during transmission, which drastically improves the throughput rate.

The network lifetime vs. a number of nodes is shown in Figure 1.4. The SC-MCHMP improved the network lifetime performance compared to the existing

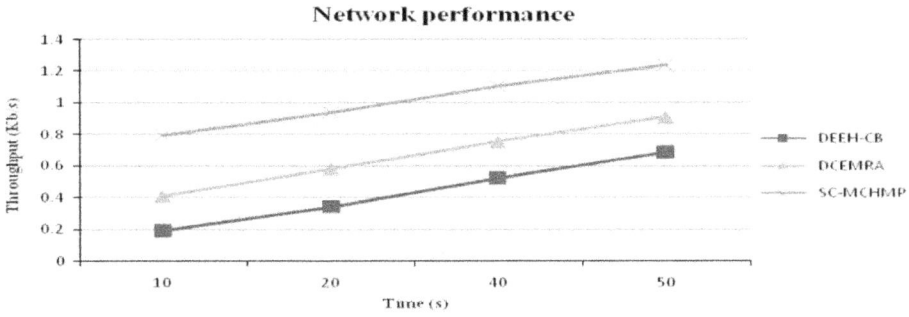

FIGURE 1.3 Network performance.

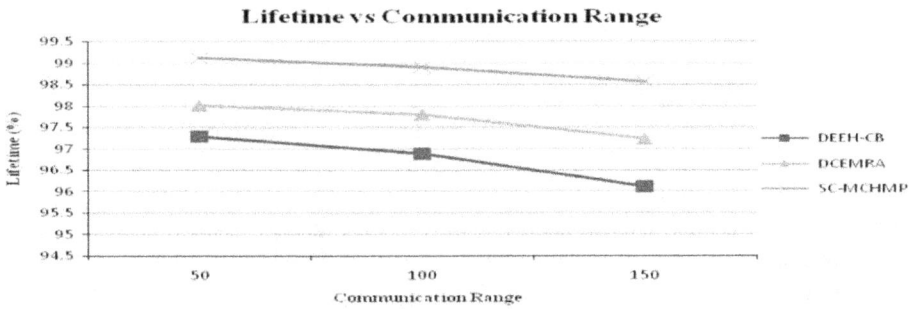

FIGURE 1.4 Network lifetime vs communication range.

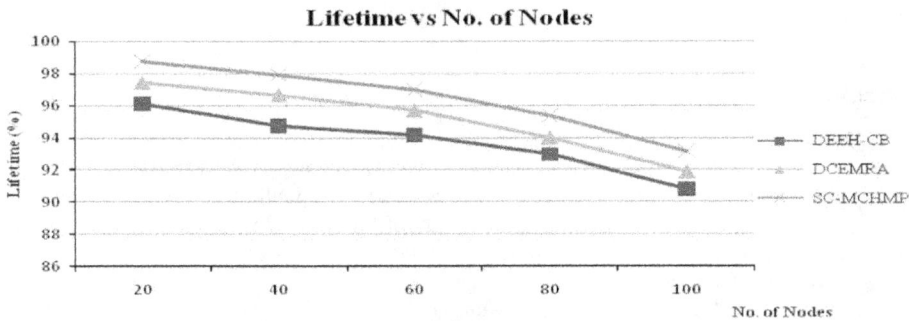

FIGURE 1.5 Network lifetime vs nodes.

methods, namely DEEEH-CB and DCEMRA due to the optimized energy consumption and the higher load balancing among the SNs. The network lifetime achieves an improvement of 13% owing to the changes in transmission and receiving nodes. Due to the reduced energy consumption ratio in SC-MCHMP, the network lifetime improves with optimized energy consumption.

The characterization of network lifetime vs. communication range is presented in Figure 1.5. The network's lifetime is enhanced using the lower communication range in this proposed method compared to the previous techniques, such as DEEEH-CB and DCEMRA. A comparable performance is possessed by this SC-MCHMP approach within the overall communication ranges. An increase of 24% is attained in the network lifetime by increasing the general communication range. Compared with the other techniques, the network lifetime increases with the rise in the distance between the nodes. The proposed SC-MCHMP method gives a similar performance in all the different communication ranges. The estimation of RSSI for score calculation provides a feasible path for all environments compared with previous protocols.

1.5   CONCLUSION

A new a novel Score based Cluster level Multi-Channel Hybrid MAC Protocol, called SC- MCHMP, is proposed by this work on behalf of the selection of routing path for the deployment of WSNs within the severe atmospheric conditions. In this method, long end-to-end delay in addition to the consumption of unbalanced energy takes place among the sensor node. The residual delivery ration RDR is calculated by this method for capturing the expected additional transmission for a unit delay. Moreover, the capability of every single sensor node for forwarding the packets is reflected by this method. The route with the best score is considered for the data transmission. The hassle-free data transmission is ensured between intra and inter clustering communication using the hybrid MAC protocols. To verify the protocol stability, this protocol has tested under various network load and dynamic network constraints. By comparing with the additional contemporary techniques, the improved network lifetime, minimized energy consumption, and enhanced network performance have been achieved using the SC-MCHMP based on hybrid MAC environment.

REFERENCES

[1] Demirkol, I., Ersoy, C., Alagoz, F. (2006). Mac protocols for wireless sensor networks: A survey. *IEEE Communications Magazine*, 44(4), 115–121.
[2] Hanjagi, A., Srihari, P., Rayamane, A.S. (2007). A public health care information system using GIS and GPS: A case study of Shiggaon. In: *GIS for Health and the Environment. Lecture Notes in Geoinformation and Cartography*. Springer, Berlin, Heidelberg, 243–255.
[3] Yang, X., Ling, W., Jia, S., Yanyun, G. (2018). Hybrid MAC protocol design for mobile wireless sensors networks. *IEEE Sensors Letters*, 2(2), 1–4.
[4] Nguyen, V., Oanh, T.T.K., Chuan, P., *et al.* (2018). A survey on adaptive multi-channel MAC protocols in VANETs using Markov models. *IEEE Access*, 6, 16493–16514.
[5] Laouid, A., Abdelnasser, D., Ahcène, B., *et al.* (2017). A distributed multi-path routing algorithm to balance energy consumption in wireless sensor networks. *Ad Hoc Networks*, 64, 53–64.
[6] Gajendran, E., Vignesh, E.J., Prabhu, B. (2017). Randomized cluster head selection strategy for dense wireless sensor networks. *International Journal of Multidisciplinary Research and Modern Education*, 3(1), 414–420.

[7] Ozdemir, S., Xiao, Y. (2009). Secure data aggregation in wireless sensor networks: A comprehensive overview, *Computer Networks*, 53, 2022–2037.

[8] Albath, J., Thakur, M., Madria, S. (2013). Energy constraint clustering algorithms for wire-less sensor networks, *Ad Hoc Networks*, 11(25), 12–25.

[9] Khediri, S.E., Nasri, N., Wei, A., Kachouri, A. (2014). A new approach for clustering in wireless sensors networks based on LEACH. *Procedia Computer Science*, 32, 1180–1185.

[10] Younis, O., Heed, F.S. (2004). A hybrid, energy-efficient, distributed clustering approach for ad hoc sensor networks. *IEEE Transactions on Mobile Computing*, 3, 366–379.

[11] Yu, J., Qi, Y., Wang, G., Gu, X. (2012). A cluster-based routing protocol for wireless sensor networks with non-uniform node distribution. *International Journal of Electronics and Communications*, 66, 54–61.

[12] Gu, X., Yu, J., Yu, D., Wang, G., Lv, Y. (2014). ECDC: An energy and coverage-aware dis-tributed clustering protocol for wireless sensor networks. *Computers & Electrical Engineering*, 40, 384–398.

[13] Wang, Z., Chen, H., Cao, Q., *et al.* (2017). Achieving location error tolerant barrier coverage for wireless sensor networks. *Computer Networks*, 112(1), 314–328.

[14] Chen, Z., Ma, M., Liu, X., Liu, A., Zhao, M. (2017). Reliability improved cooperative communications over wireless sensor networks. *Symmetry*, 9(209), 1–22.

[15] Zhou, M., Zhao, M., Liu, A., *et al.* (2017). Fast and efficient data forwarding scheme for tracking mobile target in sensor networks. *Symmetry*, 9(11), 269.

[16] Zhang, Y., He, S., Chen, J. (2016). Data gathering optimization by dynamic sensing and routing in rechargeable sensor networks. *IEEE/ACM Transactions on Networking*, 24(3), 1632–1646.

[17] Tang, J., Liu, A., Zhao, M., Wang, T. (2018). An aggregate signature based trust routing for data gathering in sensor networks. *Security and Communication Networks*. Article ID 6328504, 1–30.

[18] Wang, Z., Liao, J., Cao, Q., Qi, H., Wang, Z. (2014). Achieving k-barrier coverage in hybrid directional sensor networks. *IEEE Transactions on Mobile Computing*, 13(7), 1443–1455.

[19] Pu, L., Chen, X., Xu, J., Fu, X. (2016). D2D fogging: An energy-efficient and Incentive-aware task offloading framework via network-assisted D2D collaboration. *IEEE Journal on Selected Areas in Communications*, 34(12), 3887–3901.

[20] Hu, X., Chu, T.H.S., Leung, V.C.M., *et al.* (2015). A survey on mobile social networks: Applications, platforms, system architectures, and future research directions. *IEEE Communications Surveys & Tutorials*, 17(3), 1557–1581.

[21] Chen, X., Pu, L., Gao, L., Wu, W., Wu, D. (2017). Exploiting massive D2D collaboration for energy-efficient mobile edge computing. *IEEE Wireless Communications*, 24(4), 64–71.

[22] Dai, H., Wu, X., Xu, L., *et al.* (2015). Practical scheduling for stochastic event capture in energy harvesting sensor networks. *International Journal of Sensor Networks*, 18(1/2), 85–100.

[23] Liu, X. (2015). A deployment strategy for multiple types of requirements in wireless sensor networks. *IEEE Transactions on Cybernetics*, 45(10), 2364–2376.

[24] Dong, M., Liu, X., Qian, Z., *et al.* (2015). QoE ensured price competition model for emerging mobile networks. *IEEE Wireless Communications*, 22(4), 50–57.

[25] He, S., Gong, X., Zhang, J., Chen, J. (2014). Curve based deployment for barrier coverage in wireless sensor networks. *IEEE Transactions on Wireless Communications*, 13(2), 724–735.

[26] Liu, Y., Dong, M., Ota, K., Liu, A. (2016). Active trust: Secure and trustable routing in wireless sensor networks. *IEEE Transactions on Information Forensics and Security*, 11(9), 2013–2027.

[27] Liu, X., Dong, M., Ota, K., *et al.* (2016). Service pricing decision in cyber-physical systems: Insights from game theory. *IEEE Transactions on Services Computing*, 9(2), 186–198.

[28] Dong, M., Ota, K., Liu, A., Guo, M. (2016). Joint optimization of lifetime and transport delay under reliability constraint wireless sensor networks. *IEEE Transactions on Parallel and Distributed Systems*, 27(1), 225–236.

[29] Xu, J., Liu, A., Xiong, N., Wang, T., Zuo, Z. (2017). Integrated collaborative filtering recommendation in social cyber-physical systems. *International Journal of Distributed Sensor Networks*, 13(12), 1–17.

[30] Li, T., Liu, Y., Gao, L., Liu, A. (2017). A cooperative-based model for smart-sensing tasks in fog computing. *IEEE Access*, 5, 21296–21311.

[31] Lai, S., Ravindran, B., Cho, H. (2010). Heterogenous quorum-based wake-up scheduling in wireless sensor networks. *IEEE Transactions on Computers*, 59(11), 1562–1575.

[32] Chao, C., Lee, Y. (2010). A quorum-based energy-saving MAC protocol design for wireless sensor networks. *IEEE Transactions on Vehicular Technology*, 59(2), 813–822.

[33] Tsai, C.H., Hsu, T.W., Pan, M.S., *et al.* (2009). Cross-layer, energy-efficient design for supporting continuous queries in wireless sensor networks: A quorum-based approach. *Wireless Personal Communications*, 51(3), 411–426.

[34] Jiang, J.R. (2008). Expected quorum overlap sizes of quorum systems for asynchronous power-saving in mobile ad hoc networks. *Computer Networks*, 52(17), 3296–3306.

[35] Ekbatanifard, G.H., Monsefi, R., Yaghmaee, M.H., *et al.* (2012). Queen-MAC: A quorum-based energy-efficient medium access control protocol for wireless sensor networks. *Computer Networks*, 56(8), 2221–2236.

[36] Heinzelman, W.R., Chandrakasan, A., Balakrishnan, H. (2000). Energy-efficient communication protocol for wireless microsensor networks in system sciences. In: *Proceedings of the 33rd Annual Hawaii International Conference on System Sciences*, Maui, HI, USA, 8, 8020.

[37] Heinzelman, W.B., Chandrakasan, A.P., Balakrishnan, H. (2002). An application-specificprotocol architecture for wireless microsensor networks. *IEEE Transactions on Wireless Communications*, 1, 660–670.

[38] Chen, B., Zhang, Y., Li, Y., Hao, X., Fang, Y. (2011). A clustering algorithm of cluster-headoptimization for wireless sensor networks based on energy. *Journal of Information and Computational Science*, 11, 2129–2136.

[39] Tong, M., Tang, M. (2010). LEACH-B: An improved LEACH protocol for wireless sensornetwork. In: *Wireless Communications Networking and Mobile Computing (WiCOM), 6th International Conference*, IEEE, Chengdu, China, 1–4.

[40] Bajaber, F., Awan, I. (2011). Adaptive decentralized re-clustering protocol for wireless sensor networks. *Journal of Computer and System Science*, 77, 282–292.

[41] Hussain, S.J., Muhammad, I., Jhanjhi, N.Z., Khalid, H., Mamoona, H. (2021). Performance enhancement in wireless body area networks with secure communication. *Wireless Personal Communications,* 116(1), 1–22.

[42] Datta, D., Dhull, K., Verma, S. *UAV Environment in FANET: An Overview.* CRC Press, Taylor & Francis Group.

[43] Kumar, D., Aseri, T.C., Patel, R. (2009). EEHC: Energy efficient heterogeneous clustered scheme for wireless sensor networks.*Computer Communications*, 32, 662–667.

[44] Katiyar, V., Chand, N., Soni, S. (2010). Clustering algorithms for heterogeneous wireless sensor network: A survey. *International Journal of Applied Engineering Research Dindigul*, 1, 273–274.

[45] Bandyopadhyay, S., Coyle, E.J. (2003). An energy efficient hierarchical clustering algo-rithm for wireless sensor networks. In: *INFOCOM Twenty-Second Annual Joint Conference of the IEEE Computer and Communications, IEEE Societies*, San Francisco, CA, USA, 3, 1713–1723.

[46] Liu, Z., Zheng, Q., Xue, L., Guan, X. (2012). A distributed energy-efficient clustering algo-rithm with improved coverage in wireless sensor networks. *Future Generation Computer System*, 28, 780–790.

[47] Shahid, H., Humaira, A., Hafsa, J., Mamoona, H., Jhanjhi, N.Z., AlZain, M.A. (2021). Energy optimised security against wormhole attack in IoT-based wireless sensor network. *CMC-Computers Materials & Continua*, 68(2), 1966–1980.

[48] Vijayalakshmi, B., Ramar, K., Jhanjhi, N.Z., *et al.* (2020). An attention based deep learning model for traffic flow prediction using spatio temporal features towards sustainable smart city. *IJCS Wiley*, 34, 1–14.

[49] Guo, P., Jiang, T., Zhang, K., Chen, H.-H. (2009). Clustering algorithm in in-itialization of multi-hop wireless sensor networks. *IEEE Transactions on Wireless Communications*, 8(12), 5713–5717.

[50] Tomoya, E., Makoto, A., Takizawa, M. (2015). Energy-efficient delay time-based process allocation algorithm for heterogeneous server clusters. In: *IEEE 29th International Conference on Advanced Information Networking and Applications*, Gwangiu, South Korea, 279–286.

[51] Hsieh, M.Y. (2011). Data aggregation model using energy-efficient delay sche-duling in multi-hop hierarchical wireless sensor networks. *IET Communications*, 5(18), 2703–2711.

[52] Shafiq, M., Ashraf, H., Ullah, A., Masud, M., Muhammad, A., Jhanjhi, N.Z., Humayun, M. (2021). Robust cluster-based routing protocol for iot-assisted smart devices in WSN. CMC-Computers Materials & Continua 67(3), 3505–3521.

[53] Sowjanya, R., Kavita, Verma, S. (2019). The amalgamative sharp WSN routing and with enhanced machine learning. *Journal of Computational and Theoretical Nanoscience (JCTN)*, 16(9), 3766–3769.

[54] Shokouhifar, M., Jalali, A. (2017). Optimized sugeno fuzzy clustering algorithm for wireless sensor networks. *Engineering Applications of Artificial Intelligence*, 60, 16–25.

[55] Batra, I., Verma, S., Kavita, Mamoun, A. (2019). A lightweight IoT based security framework for inventory automation using wireless sensor network. *IJCS Wiley*, 33, 1–16.

[56] Batra, I., Verma, S., Kavita, *et al.* (2020). Hybrid logical security framework for privacy preservation in the green internet of things. *MDPI-Sustainability*, 12(14), 5542.

[57] Jan, M.A., Dong, B., Jan, S.R.U., *et al.* (2021). A comprehensive survey on machine learning-based big data analytics for IoT-enabled smart healthcare system. *MONET, Springer*,

[58] Huang, R., Zhu, J., Yu, X.-T. (2006). The Ant-based Algorithm for the data gathering routing structure in sensor networks. In: *Proceedings of the Fifth International Conference on Machine Learning and Cybernetics*, Dalian, China, 4473–4478.

[59] Heinzelman, W.R., Chandrakasan, A., Balakrishnan, H. (2000). Energyefficient communication protocol for wireless microsensor networks. In: *Proceedings of IEEE 33rd Annual Hawaii International Conference on System Sciences*, Maui, Hawaii, USA, 1–10.

[60] Sabet, M., Naji, H.R. (2015). A decentralized energy efficient hierarchical cluster-based routing algorithm for wireless sensor networks. *AEU-International Journal of Electronics and Communications*, 69(5), 790–799.

[61] Singh, P., Verma, S., Kavita. (2019). Analysis on different strategies used in blockchain technology. *Journal of Computational and Theoretical Nanoscience (JCTN)*, 16(10), 4350–4355.

[62] Ghosh, G., Kavita, Verma, S., Jhanjhi, N.Z. (2020). Secure surveillance system using chaotic image encryption technique. *IOP Science*, 993, 012062.

[63] Srivastava, A., Verma, S., Kavita, Jhanjhi, N.Z., Malhotra, A. (2020). *Analysis of Quality of Service in VANET*. IOP Science.

[64] Selvi, M., Velvizhy, P., Ganapathy, S., Khanna, N.H., Kannan, A. (2019). A rule based delay constrained energy efficient routing technique for wireless sensor networks. *Cluster Computing*, 22(5), 10839–10848.

[65] Saeed, S., Jhanjhi, N.Z., Naqvi, M., Humayun, M., Ponnusamy, V. (2020). Analyzing the performance and efficiency of it-compliant audit module using clustering methods. In: *Industrial Internet of Things and Cyber-Physical Systems: Transforming the Conventional to Digital*, IGI Global, 351–376.

[66] Humayun, M., Jhanjhi, N., Alruwaili, M., Amalathas, S.S., Balasubramanian, V., Selvaraj, B. (2020). Privacy protection and energy optimization for 5G-aided industrial internet of things. *IEEE Access*, 8, 183665–183677.

2 Software-Defined Networking (SDN) Security Concerns

*Muhammad Junaid Nazar, Saleem Iqbal,
Saud Altaf, Kashif Naseer Qureshi,
Khalid Hussain Usmani, and Sobia Wassan*

CONTENTS

2.1   INTRODUCTION

Software-Defined Networking (SDN) is a networking approach to enable network administrators to respond rapidly through (logically) a centralized controller. It may change the limitations of classical network infrastructure. SDN broke the vertical integration and separates the control plane (controller) from the data plane (switches) that forward the traffic. The main idea behind SDN is to decouple the data plane/forwarding plane from the control plane, where the controller controls and manages the device that resides in the forwarding plane. The centrally controlled control plane benefits the SDN because it's flexible and programmable and

DOI: 10.1201/9780367808228-2

19

has control over network traffic flows. The control and data plane are decoupled, which could be understood using a programming interface among SDN controller and switches. The network administrator can manage and control the network and packet-processing functions through his own program.

In a classical network, there are different algorithms implemented on hardware devices to monitor and control the flow of data, manage routing paths, and determine the connection of different devices in the network. Generally, the defined rules and algorithms to route the network traffic are applied in hardware components such as Application Specific Integrated Circuits (ASIC) [1]. An essential significance of the SDN principle is the concern for the separation among network policies and the implementation of these policies in traffic forwarding and hardware. This dissociation is essential to flexibility, breaks the network into controllable pieces, creates and introduces new abstraction in networking, and simplifies and facilitates the network management, innovation, and evolution [1].

The isolation of data and control plane between the controller and switches can be realized by the well-defined interface of programming. The data plane devices are controlled by the SDN controller via the well-defined application programming interface (API). The noteworthy example of an API is OpenFlow [1,2]. An OpenFlow is a protocol that is used for communication between controllers and switches. OpenFlow-enabled switches consist of a table called a flow table that handles packet-forwarding rules and provides a secure channel for communication. In the flow table, each flow rule matches the flow entry and performs assured actions, i.e., forward the flows to destination or controller, drop the packet, modify traffic etc. The controller instructs the OpenFlow-enabled switch to behave like a switch, and the router performs roles like traffic shaping, load balancing etc., while OpenFlow and SDN begin as academic experiments [3].

With the current scenario of network-devices implementation, the flow table plays an important role in SDN. Each OpenFlow switch has a data path that is calculated by the controller, and the flow table consists of predetermined set of rules of network flows. The controller generates the flow rule and installs them to the flow table, which consists of three portions: (1) matching pattern; (2) action associated with matching pattern, and (3) statistics. The statistics portion consists of priority, idle timeout, hard timeout, packet, and byte counter, etc. Among various attacks, the overflow attack targets the flow tables of forwarding devices in the data plane to overflow the flow tables and drop the packets. If the flow table is compromised by a malicious attacker, its performance is affected by the malfunctioning flow table, such as manipulation and conflicts in defined rule and exhaustion of resources. These flow rules are generated or set by the controller using either proactive or reactive approaches [4].

An architecture of SDN is presented in Figure 2.1. It is necessary to emphasize a programmatic model that does not assume to be centralized physically, though the system is (logically) centralized [5]. In fact, it's important to ensure sufficient levels of adaptability, performance, reliability, and scalability. The generation level of SDN organizes outlines to physically distribute the control plane [5,6].

A software-defined network has anomalies and security issues that disturb the behavior of data flow; the problem is addressed by securing the controller and

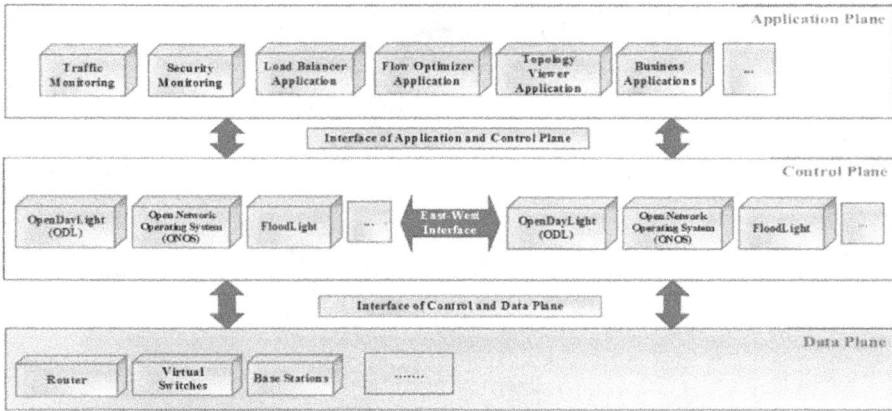

FIGURE 2.1 Architecture of software-defined networking.

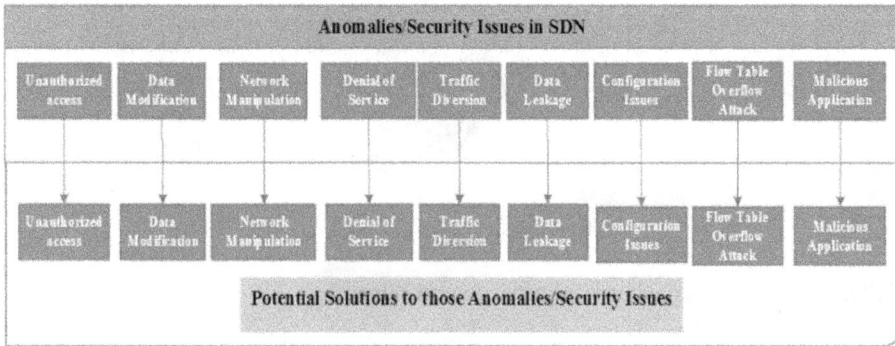

FIGURE 2.2 Overview of anomalies in software-defined networking.

switches. The focus is on different anomalies, including a flow-table overflow attack that affects the whole network. The aim is to present the definition and architecture, security, and solutions of security issues in SDN, provide an overview of recent research and development, and define how an attack can create disruption in network devices. Figure 2.2 illustrates the overview of SDN security and indicates the category of anomalies, their solution, and the enhancement of security in SDN using SDN framework.

Figure 2.3 representing the particular characteristics of an architecture of SDN that might have an effect on the security of SDN through presenting the vulnerabilities or enhancing the network security. The six features of SDN are illustrated in Figure 2.3; there is a centralized controller (logically), network hypervisor, switch management cluster, network services, virtualized network, and centralized monitoring unit. The instance of each specific controller is the master or primary controller of a controller set, and forwarding devices are bunched in master and slave manner. The network services include business applications, such as routing,

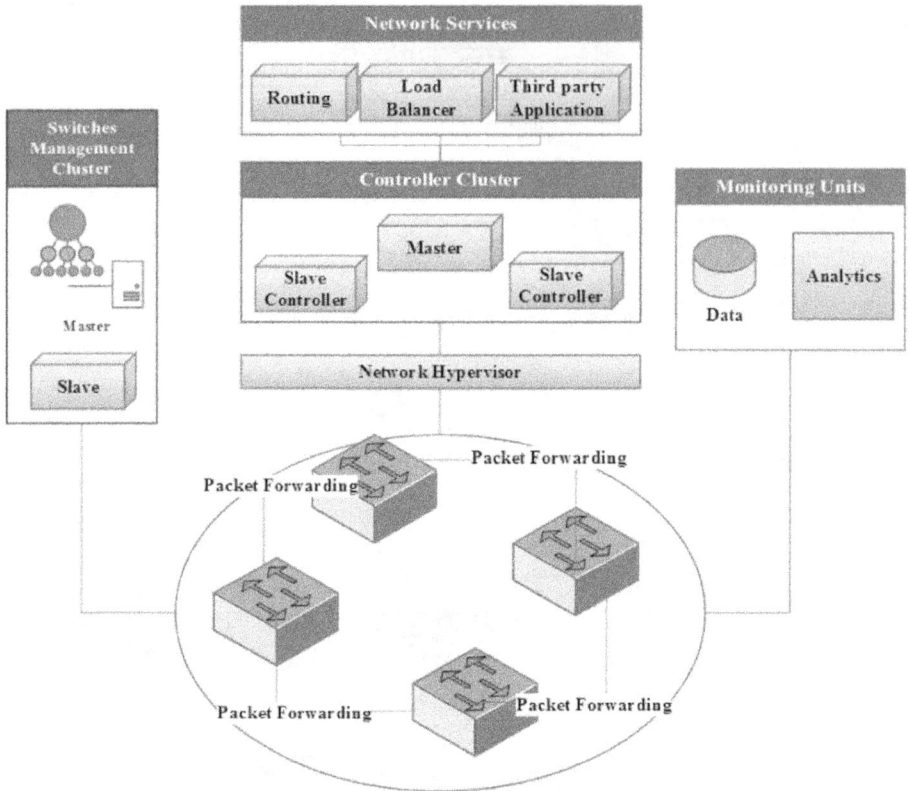

FIGURE 2.3   Features of software-defined networking.

load balancer and third-party application for network management, topology viewer, flow optimizer applications, etc. The controller and switch management clusters are included in a master and slave manner. The monitoring unit includes a database and analytics portion in which statistics are maintained, monitored, and also saved in the database. The remaining two features include a network hypervisor, an implementation layer and open-flow enabled switches.

The purpose to model the anomalies in SDN is to identify those that originate at a data plane. The compromised switch might have random behavior, i.e., modifying packets, dropping traffic, or diverting traffic from another path rather than a defined path. It might send wrong statistics reports to the controller, collected from switches to hide the malicious behavior. In fact, this misbehavior occurs by any malicious activity, which forwards packet reports and packets to the wrong destination via the wrong path. The malicious attackers can make the misbehavior in switches and take control over the network. Forwarding anomalies could be from failure of network equipment or hardware, misconfiguration, or software bugs. Figure 2.4 shows the tree representation of various anomalies and security issues that affect the SDN application, control, and data plane. Denial of Service (DoS) is a common attack

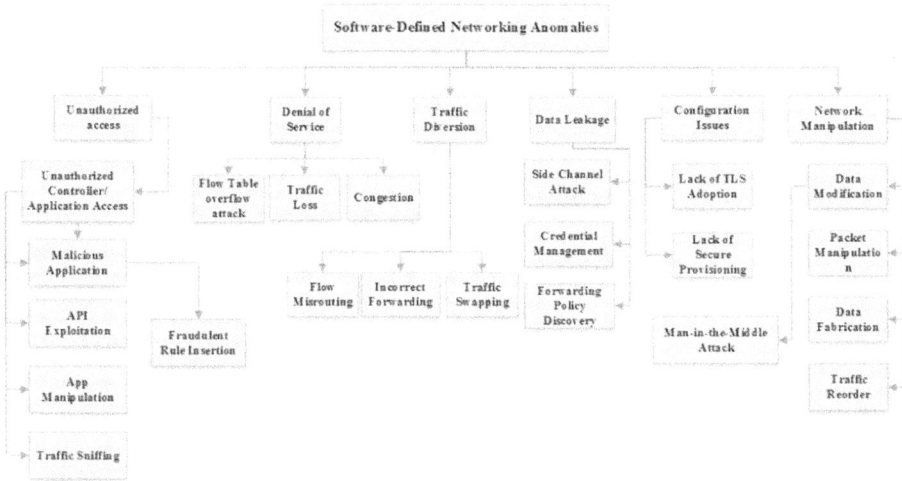

FIGURE 2.4   Tree representation of SDN attacks and security issues.

that affects classical networks, as well as all layers of SDN. A flow-table flooding attack is one type of DoS attack that drops the packets as well as diverts network traffic after injecting malicious flow rules and makes switches to take other paths rather than an original path.

2.2   POTENTIAL ANOMALIES IN SDN

Some anomalies and security issues related to SDN are acknowledged in Figure 2.5. The study associated with the solutions is categorizing and presenting in the literature shown in Table 2.1 and comparing the solutions in contradiction of the appropriate planes of SDN.

2.2.1   UNAUTHORIZED ACCESS

Unauthorized access is attributed to either controller access or unauthenticated application. A centralized controller is one of the leading characteristics of SDN. The issue is to reconfigure the network or elements of SDN framework damaged by any unauthorized entity. Different types of anomalies have been defined in various studies regarding unauthorized access, including malicious application, API exploitation, APP manipulation and traffic sniffing. The abstraction is provided by the controller to the application so that applications can accomplish the reading/writing of the network state. The above-mentioned four types of anomalies can compromise the application for gaining access to the resources of the network to manipulate, sniff, and exploit the operation of the whole network. To improve the effectiveness of the network and remove the bottleneck of the controller, the authors of [7] described a hybrid control model that secures the distributed control model. In [7], the control of the network is centralized under usual conditions and circumstances, but if the network has a heavy load, the equipment of network installs the flow rules on

TABLE 2.1

Research Work Comparison on Solutions to Various Anomalies in SDN

Research Work	SDN Planes					Anomalies in SDN					
	Application Plane	Northbound Interface	Control Plane	Southbound Interface	Data Plane	Unauthorized Access (i.e., controller/ Application)	Data Leakage	Network Manipulation	Traffic Diversion	DoS (i.e., Flow Table Flooding, etc.)	Configuration Issues
FSL (2008) [28]	✓		✓		✓						✓
Ristenpart et al. (2009) [29]					✓		✓				
FlowChecker (2010) [23]	✓		✓		✓						✓
VAVE (2011) [30]			✓		✓					✓	
ProtoGENI (2011) [31]			✓		✓	✓		✓			
CPRecovery (2012) [32]			✓		✓					✓	
FortNOX (2012) [9]	✓		✓			✓					
VeriFlow (2012) [26]	✓		✓	✓	✓						✓
NICE (2012) [19]	✓		✓		✓						
Kruetz et al. (2013) [33]			✓		✓			✓		✓	✓

STRIDE and Attack Modelling Method (2013) [34]	✓		✓		✓		✓			
Benton et al. (2013) [18]	✓		✓		✓		✓	✓		
Shin et al. (2013) [35]	✓		✓			✓		✓		
Othman et al. (2013) [7]	✓		✓		✓		✓	✓		✓
Yu at al. (2013) [36]			✓		✓			✓		✓
PermOF (2013) [37]	✓		✓	✓	✓		✓	✓		✓
AVANT-GUARD (2013) [38]							✓			
NetPlumber (2013) [27]							✓			
FLOVER (2013) [24]							✓			
Wang et al. (2013) [39]	✓						✓		✓	✓
Smeliansky (2014) [40]										
Ding et al. (2014) [41]	✓				✓		✓		✓	✓

(Continued)

26

Information Security Handbook

TABLE 2.1 (Continued)
Research Work Comparison on Solutions to Various Anomalies in SDN

AuthFlow (2016)[********]	✓							✓		✓	✓	
Li et al. (2014)[42]					✓	✓			✓			
Payless (2014)[14]									✓			
Operation Checkpoint (2014)[43]	✓		✓					✓			✓	
OpenSample (2014)[22]							✓	✓				
Entropy based (2014)[44]				✓	✓			✓				
ROSEMARY (2014)[10]	✓						✓	✓				
LegoSDN (2014)[11]	✓							✓				
FLOWGUARD (2014)[45]	✓						✓	✓	✓			
FlowVisor (2014)[46]						✓			✓			
Planck (2014)[12]												
SE-Floodlight (2015)[47]	✓							✓	✓			

FloodGuard (2015) [48]

Tommy et al. (2015) [49]

FlowRanger (2015) [50]

Chi et al. (2015) [51]

SPHINX (2015) [15]

FlowMon (2015) [52]

AuthFlow (2016) [53]

Qian et al. (2016) [4]

LineSwitch (2016) [54]

Aleroud et al. (2016) [55]

FADE (2016) [56]

FTGuard. (2017) [57]

Xu et al. (2017) [58]

Zhou et al. (2018) [59]

behalf of the controller to other network equipment. The authors present the signature algorithm for secure transmission of flow requests from the network device to another device.

2.2.2 MALICIOUS APPLICATION

The controller acts as an abstraction for applications from the data plane because there is an abstraction between control and data plane. The third-party application has been enabled by SDN and incorporated in the architecture [8]. In the same way, the less efficient delineated application might present vulnerabilities involuntarily. The solution of this type of anomaly is the establishment of a trusted connection between applications and controllers for the secure exchanging of control messages. More than a few solutions are presented in [9–11]. In [9], the authors proposed a security-enforcement kernel relating to this security issue. The role-based authentication system is implemented by FortNOX determine the authorization of security of each application. FortNOX handles the possible conflicts with the insertion of flow rule, whereby acceptance and rejection are reliant on security authorization. In [10], the secure, robust, and high performance net operating system (NOS) has been proposed. The main intent of [11] is to increase the flexibility of controller to malicious applications. In [11], the isolation layer is proposed for avoiding the crash of the application caused the collapse of controller. This isolation layer among the applications, a transaction system, and the layer of fault tolerance identifies and overwhelms the crash-generating actions.

2.2.3 NETWORK MANIPULATION

Network manipulation takes place in the control plane. The SDN controller compromised by an attacker to produce incorrect data initiates other attacks on the entire network. Different researches have used different terminologies of network manipulation. Traffic modification, packet manipulation, and traffic fabrication are types of network manipulation. The switches are maliciously compromised to modify the packet/traffic content. The relation between network manipulation with its types are: the controller has been compromised by an attacker to produce wrong data, initiates other attacks on the data plane, and those attacks could be traffic modification, packet manipulation, and traffic fabrication.

1. *Data modification* is the alteration in data traffic forwarded from source to destination. In SDN, the modification can occur in the network by the malicious switch. The content of traffic could be changed by malicious switches, by either payload or overhead of packets. The possible solution to this type of attacks are security techniques like cryptographic techniques; for instance, the one-way hash function guarantees the reliability of data traffic, but in SDN, the detection system for the controller is more useful than those mechanisms. For instance, with the traffic mirroring in Planck [12] and the data traffic forwarded to the collector, the veracity of entire packets could be checked on various switches. The topology and flow rules are shared by the controller to the collector so that the collector might identify the input and output ports of the flow. It calculates

the packet digestion at every switch in the network and sends those packets to
the detection system for the purpose of checking the integrity. The detection
system compares those digested packets among different switches to avoid
false negatives, for example, the time-to-live and checksum.

i. ***Man-In-The-Middle Attack*** is a type of data modification. It occurs when
an attacker has the ability to infiltrate the whole network using
ARP Spoofing. The attacker can sniff, modify, and even stop the data
traffic. This could happen when there is no authentication mechanism on
endpoints of network or communication. A network hypervisor, LiteVisor
[13] for OpenFlow, was recognized to lack an appropriate isolation
technique [14], which would permit an attacker to initiate the data-
modification attack.

2. ***Traffic Fabrication*** is when the arbitrary packets could be generated by malicious switches and then these packets could be forwarded to the data plane or
on the control plane. A possible solution of traffic fabrication is flow preservation that can be used in the data plane, but it would not detect the packets
dropping switches and insert the same number of bogus packets. To be more
effective, this solution could be combined with improved traffic detection. The
framework Sphinx [15] allows the administrator to enforce the constraints on
flows and validate network-wide invariants. An example is bogus ARP messages identifying the Sphinx and maintaining a list of MAC address to IP
binding. That bogus traffic might be used for DoS attack in the control plane. In
this case, rate-limiting schemes could be used to control the number of requests.

3. ***Packet Manipulation*** is when the packet is modified by an attacker
that passes the compromised switch. The packet manipulation can block
communication using decryption errors; however, it might protect by any
encryption technique.

4. ***Flow Reorder*** means that, generally, the traffic order might change by malicious switches but still be effective in terms of routers, delay, and content.
With the aspect of delay in network traffic, uninterrupted retransmission of
TCP packets would degrade the throughput [16]. The probable solution of this
type of anomaly is similar to the traffic modification. The packets order would
be checked by a detection system for keeping an ordered list of packets
digested from each switch.

2.2.4 TRAFFIC DIVERSION

Traffic diversion is a data plane attack in which the attack compromises all network
elements to divert or redirect traffic flows, allowing eavesdropping in addition. The
solution of traffic diversion attack is to use strong encryption to secure the network
elements and the network's channels of communication. Traffic misrouting, incorrect
forwarding, and packet swapping is also a traffic diversion attack with different
naming conventions that is used in research work. The switches might be compromised by an attacker to command the packets to progress without ensuring the set flow
rules. As an illustration, the OpenFlow (OF) switch is programmed by the controller to
forward the packets from port 1 to 3, but an attacker modifies the program and

commands to forward the packet to port 4. It will interrupt the entire forwarding and switching system. This misbehavior could be resolved by applying a reactive approach rather than a proactive approach. In such case, when a switch receives a flow, it gets instructed by the controller by sending specified commands and its response regarding the flow table. However, if the switches had previously populated flow rules, including expired ones, the switch would not generate messages to the controller. This would break the entire switching system and might create infinite loops. Those terminologies lead to the same attack and are also described in Table 2.1.

2.2.5 DENIAL OF SERVICE (DOS)

This is the most common type of attack that can affect all planes of SDN. DoS could cause disruption or reduction of SDN services. The solution of a DoS attack is to apply packet-dropping techniques and rate limiting at the control plane. As the communication path between the controller and forwarding devices, the controller could be flooded with packets populated by an attacker wanting to take complete control of the flow rule and make it unapproachable to authentic users. The attack can also be accomplished at the data plane with the flow table flooding, with the result being that the availability of memory is limited [17]. In addition, the fraudulent rule insertion and modification have also led by DoS discussed in [18]. Switch flow table flooding is also called flow table overflow attack, and controller-switches communication flooding is described in DoS. Data traffic loss and delay has also performed by DoS, when an attacker compromises and floods the controller and switches, and the switches could drop packets randomly.

1. **Overflow Attack** is a switch-flow table flooding attack type of DoS that leads to making both controllers and switches vulnerable. The switches could not take more entries of flow when flow table is overflowed, and the controller would not reply to the requests of the appropriate client. The installation of new rules causes packet loss because it involves enervation of flow entries that make flow table overflow [4]. The flow table has overflowed either by attack from a malicious application on controller or attack from packets.
    i. **Attack on control plane as of Malicious Application:** The controller is the main part of the SDN network and takes the decision of network packets and flow entries that have match or no match in the flow table. The flow table has new rules for flow installed by the controller for each packet or flow. The malicious app would make an attack after deploying in the controller to take care of all messages. The solution of this type of attack is to assess the application sensibly before installing a controller. Kostic et al. [19] anticipated a means for testing applications. When messages are received from the switch, the malicious application starts an infinite loop for installing multiple flow rules. The large number of new rules could insert and permute the source IP and destination IP to overflow the flow table.
    ii. **Attacks from packets:** Attacks are internal as well as external, and attacks originated by external attackers have elastic methods to launch a DoS

attack. The attacker could generate and send bulk packets to the controller, resulting in the installation of new rules for respective packets and exhaust the flow table. The fields of source and destination port are permuting in the header of the packet to craft the number of UDP packets in Scapy [20] and send flow packets at a user-defined rate.

2. *Data Loss*: The data traffic could drop randomly by malicious switches in a selective way. This could happen by any attack occurring in data plane switches or over the path. In packet loss, the statistics data of every switch is useful on a path of flow and apply the flow-conservation principle in which incoming traffic flow must be equivalent to outgoing flow. This principle is applied on per flow and is given the finest accuracy; in this way, we might detect the lost traffic exactly. However, two difficulties are met: 1) valid packet loss occurs because of congestion, and 2) counter synchronization occurs between dissimilar switches. The significance to the valid packet loss is accurate measure of congestion. Congestion can cause the packet loss, and if the number of lost packets is known, then it eliminates the vagueness; over this number, any lost packet is considered defective. If the purpose is to attain the loss detection of real-time traffic, then the evaluation of the number of lost packets is not a direct task. Secondly, relating counters imposes by taking a reliable global snapshot of counters on respective switch. The packets received for each flow are well-known implementation in OpenFlow of SDN [21]. The controller uses the stats_request message of OpenFlow for the query of the flow statistics and the switch used stats_reply message for the query of the number of bytes and received packets contained in each flow. In [22], the control loop of 100 millisecond is achieving real-time detection of congestion.

3. *Traffic Delay:* The malicious activities in a network might increase in jitter and delay in the network traffic, which would pose an utmost challenge to time-sensitive data traffic. Moreover, the delay in TCP traffic stream would cause false timeout and unnecessary retransmissions and affect the throughput [9]. There is a probable solution to monitor the delay in the network, provide the real time delay, and compare the arrival time of packets among switches for resolving this issue. Chowdhury et al. [14] proposed the PayLess, a framework of monitoring for SDN that collects flow statistics and provides to controller and different levels of collection. It shows tradeoffs between the monitoring accuracy and overhead of network.

2.2.6 CONFIGURATION ISSUES

For the detection of network anomalies/vulnerabilities, protocols and policies related to network security are continuously developed. Many policies and protocols will apply to the planes and architecture of SDN. In SDN, it is, and it will be, essential to impose the implementation of developed policies such as transport layer security (TLS). The interface between different components of the network has the possibility to present significant vulnerabilities about interoperability between devices of multiple vendor and communication of control/data across new interfaces. SDN provides the creation of flow rules and policies dynamically, as well as the ability to program the network using API. Quite a few solutions have been proposed to the issues relating to a conflict of network policy from multiple applications in the [19,23–27]. NICE [19] is proposed, in which the OF application is tested by combining the model checking with symbolic execution. The binary decision diagram has been proposed in [23] for testing of intra-switch misconfigurations within particular flow tables. Similarly, [24] uses modulo theories and assertion sets for the purpose of verifying policies of flow. In [25], authors proposed the approach for diagnosing configuration problems of the network using static analysis. [26] has detected the loops in the routing table, paths, etc., by modeling the graph and studying the invariants by interrupting flow rules before reaching the network. [27] is also a tool for checking the policies in real time that gradually checks the changing state. Further deployments of experiments could estimate the bond and the gap among network characteristics and theoretical models.

2.2.7 DATA LEAKAGE

For packet handling, there are diversifications of possible actions explained in OpenFlow switch specification [2]. Those actions include packet dropping, forwarding, and sending to the controller. By means of packet-processing timing analysis, an attacker decides the action to apply on specific types of packets. For example, network packet entry and out time would be shorter than a processing time on a controller. This can be done by discovering the proactive and reactive configuration of switches by an attacker. Another open challenge in SDN is storage of credentials that must be secure, e.g., certificates and keys for multiple logical networks. Side-channel attacks and a security mechanism credential management are described in data leakage. Side-channel attack occurs on the data plane of SDN. The attacker can compromise the switches and gain secure information of implementation of any system or from targeted resources. Multiple OpenFlow logical switches are initiated on top of OF enabled switch by OF-Config. Different customers have assigned logical entity, and those logical networks are linked with credentials, e.g., certificates and keys. If those credentials are not securely contained, then this can be compromised by attacker and lead to data leakage. The solution is to use a strong encryption algorithm to secure the elements and resources of network.

Table 2.2 illustrate the inappropriate use of security features that can affect all planes of SDN architecture. The table shows categorization of SDN attacks that is associated with SDN planes.

2.3 SECURITY ANALYSES AND RESEARCH CHALLENGES

2.3.1 SDN SECURITY ANALYSES

Several SDN security analyses lead this study [18,31,33–35,40,41,46,60–62]. This literature has established the relationship between the elements in framework of SDN in order to introduce new vulnerabilities. However, the recent analyses identify the enhancement of security in SDN. We know that OpenFlow is the

TABLE 2.2

Categorization of Attacks Associated with SDN Planes

SDN Security Attacks	Affected SDN Planes				
	Application Plane	App-Ctrl Interface	Control Plane	Ctrl-Data Interface	Data Plane
Data Modification			✓	✓	✓
Unauthorized controller access			✓	✓	✓
Unauthorized Application	✓	✓	✓		
Side Channel Attack					✓
Network Manipulation			✓		
Traffic Diversion					✓
API Exploitation	✓	✓			
Credential Management					✓
Flow Table Overflow Attack					✓
Malicious Application	✓	✓	✓		
Configuration Issues (Lack of secure Provisioning and TLS)	✓	✓	✓	✓	✓
Traffic Sniffing			✓	✓	✓

protocol and common technology of SDN, and [34] an analysis of OF protocol is completed by STRIDE threat exploration [63].

The described security analyses are relatively presented in Table 2.1 with respect of SDN layers and their solutions provided in different research work. The OpenFlow and SDN framework support application to modify the state of network in switches through protocol. As a result, the issues related to control and data plane, and the data-control interface, while rare [1,33,60], highlight the security threats of application hosting by third parties.

2.3.2 SCENARIOS

2.3.2.1 Scenario 1

The network flow is being diverted by malicious application. As it found, the network flow might change the path to reach the destination. It can happen by fraudulent rules insertion through malicious application. OF-enabled switches forward the whole traffic using flow rules defined by the controller. The attacker compromised the controller to access the application, making the existing application malicious by adding new rules or modifying existing rules. By those modified rules, flow entries could change in the flow table and direction would be changed due to attack.

2.3.2.2 Scenario 2

In the analysis of traffic diversion in the SDN forwarding plane, the network flow traffic can also change its path due to link failure. In the data plane, when the link is a failure or faulty, the OF-enabled switches change the path for flow arriving from source. Switches can fail over the flow that is affected due to any reason or link failure to another path. The controller calculates the multiple paths and pre-establishes them; when link failure occurs, it can take a new path to reach destination. This is general failure and take network traffic to other path.

2.3.2.3 Scenario 3

In case of congestion, it occurs when the link is carrying too much data when arriving from controller. This can cause network traffic loss and reduction of throughput, and link congestion occurred due to DoS attack. The attacker sends too much malicious traffic to overload the OF-enabled switches. In SDN, each switch records the path load information of the outgoing link. If the controller calculates the paths using a proactive approach and congestion occurs on links, the typical effect will be packet loss, but packets might change the path due to a pre-calculated/pre-established path by the controller to reach the destination.

2.3.3 RESEARCH CHALLENGES

In an SDN rather than a classical network, network traffic management is a very important subject for the optimization of the network and its performance by dynamically analyzing, predicting, and normalizing the propagated data behavior. Highly effective network management requires improved utilization of resources for optimal performance of the system when it comes across the demand for large-scale data centers. There are few open research challenges and issues regarding SDN, highlighted here.

a. Cost efficient and fast failure recovery in an SDN forwarding plane accomplished with less interference and low overhead of communication.
b. Manage big data efficiently, analyze the network traffic in the perspective of time dependent statistics, user behavior, and locality.
c. Monitor network traffic to identify how to decrease the network overhead while network statistics are collected by controller.
d. Efficiently update and control the network information in large-scale SDN network with reliability in the existence of packet loss using a single controller.
e. Reliably update the information of distributed network in whole network with real-time update and inter-synchronization overhead using multiple controllers.
f. Backup controllers in the case of failure of primary/master controller for achieving robustness.
g. Dynamic load balancing of data and control plane to utilize the flexible global and control view and avoiding the bottleneck at single centralized controller. The load balancing enables accurate and efficient acquisition of network traffic statistics.

2.4 CONCLUSION

In this chapter, the network security in SDN has been categorized as promising to improve using the features of its architecture. The vulnerabilities and little enhancements to the network security through SDN are described; they are considered more mature than in a classical network. Nevertheless, various research problems are highlighted, and several solutions have been presented related to network security attacks, specifically damaging the SDN. This chapter specifically targets the SDN security concerns and anomalies in which packets are dropped due to any malicious activity by an attacker, i.e., congestion, attack, or link failure, and, sometimes, packets change the egress port or take their own direction to reach the destination.

The type of DoS attack is also identified in which the flooding of packets causes flow-table overflow. The identified traffic diversion is an attack that compromises the switch by changing flow rules and entries in a flow table done on the controller [64]. Having research on SDN security, different approaches are used to tackle the security issues and anomalies in SDN and compare the research work on solution, as well as categorize the working and targeted SDN plane in the table above. The software-defined networking may perhaps be more secure than a classical network by implementing the techniques of security from network deployment and exploiting the programmable and dynamic SDN characteristics. For efficient detection of attacks, further improvement in SDN-based network security need to be considered in future research goals in which traffic-diversion attack and parameters of port scanning ought to be explored.

REFERENCES

[1] Bera, S., Misra, S., & Vasilakos, A. V. (2017). Software-defined networking for internet of things: A survey. *IEEE Internet of Things Journal*, 4(6), 1994–2008.

[2] Open Network Foundation. [Online]. Available: https://www.opennetworking.org/.

[3] Al-Tam, F., & Correia, N. (2019). On load balancing via switch migration in software-defined networking. *IEEE Access*, 7, 95998–96010.

[4] Alayda, S. (2021). Terrorism on Dark Web. *Turkish Journal of Computer and Mathematics Education (TURCOMAT)*, 12(10), 3000–3005.

[5] Almrezeq, N. (2021). Cyber security attacks and challenges in Saudi Arabia during COVID-19. *Turkish Journal of Computer and Mathematics Education (TURCOMAT)*, 12(10), 2982–2991.

[6] Fawcett, L., Scott-Hayward, S., Broadbent, M., Wright, A., & Race, N. (2018). Tennison: A distributed SDN framework for scalable network security. *IEEE Journal on Selected Areas in Communications*, 36(12), 2805–2818.

[7] Champagne, S., Makanju, T., Yao, C., Zincir-Heywood, N., & Heywood, M. (2018, July). A genetic algorithm for dynamic controller placement in software defined networking. In Proceedings of the Genetic and Evolutionary Computation Conference Companion (pp. 1632–1639).

[8] SDN Dev Center: Unlock networking innovation. [Online]. Available: http://h1 7007.www1.hpe.com/in/en/networking/solutions/technology/sdn/devcenter/index.aspx#. WttQjYhubIU.

[9] Birkinshaw, C., Rouka, E., & Vassilakis, V. G. (2019). Implementing an intrusion detection and prevention system using software-defined networking: Defending against port-scanning and denial-of-service attacks. *Journal of Network and Computer Applications*, 136, 71–85.

[10] Akin, E., & Korkmaz, T. (2019). Comparison of routing algorithms with static and dynamic link cost in software defined networking (SDN). *IEEE Access*, 7, 148629–148644.

[11] Humayun, M., Jhanjhi, N. Z., Alruwaili, M., Amalathas, S. S., Balasubramanian, V., & Selvaraj, B. (2020). Privacy protection and energy optimization for 5G-aided industrial internet of things. *IEEE Access,* 8, 183665–183677.

[12] Humayun, M., Niazi, M., Jhanjhi, N. Z., Mohammad, A., & Mahmood, S. (2020). Cyber security threats and vulnerabilities: A systematic mapping study. *Arabian Journal for Science and Engineering,* 45(4), 3171–3189.

[13] Yang, G., Yu, B. Y., Kim, S. M., & Yoo, C. (2018). LiteVisor: A network hypervisor to support flow aggregation and seamless network reconfiguration for VM migration in virtualized software-defined networks. *IEEE Access*, 6, 65945–65959.

[14] Chowdhury, S. R., Bari, M. F., Ahmed, R., & Boutaba, R. (2014). PayLess: A low cost network monitoring framework for Software Defined Networks. In Proceedings of the *IEEE/IFIP Network Operations and Management Symposium* (pp. 1–9).

[15] DhawanM. (2015). SPHINX: Detecting security attacks in software-defined networks. *Ndss '15*, 8–11.

[16] Aujla, G. S., Chaudhary, R., Kumar, N., Kumar, R., & Rodrigues, J. J. (2018, May). An ensembled scheme for QoS-aware traffic flow management in software defined networks. In Proceedings of the 2018 IEEE International Conference on Communications (ICC) (pp. 1–7). IEEE.

[17] Latah, M., & Toker, L. (2018). A novel intelligent approach for detecting DoS flooding attacks in software-defined networks. *International Journal of Advances in Intelligent Informatics*, 4, 2018.

[18] Zhao, C., & Liu, F. (2018). DDoS attack detection based on self-organizing mapping network in software defined networking. In *Proceedings of the MATEC Web of Conferences* (p. 01026). EDP Sciences.

[19] Canini, M., Venzano, D., & Pereˇ, P. A NICE way to test OpenFlow applications.

[20] Scapy. [Online]. Available: http://scapy.readthedocs.io/en/latest/.

[21] Protocol, V. (2014). *OpenFlow switch specification*. Open Networking Foundation.

[22] Suh, J., Taekyoung, T., Dixon, C., Felter, W., & Carter, J. (2014). OpenSample: A low-latency, sampling-based measurement platform for commodity SDN. In *Proceedings of the 2014 IEEE 34th International Conference on Distributed Computing Systems* (pp. 228–237).

[23] Han, Y., Rubinstein, B. I., Abraham, T., Alpcan, T., De Vel, O., Erfani, S., ... & Montague, P. (2018). Reinforcement learning for autonomous defence in software-defined networking. In *International Conference on Decision and Game Theory for Security* (pp. 145–165). Springer, Cham.

[24] Son, S., & Porras, P. Model checking invariant security properties in OpenFlow.

[25] Yao, H., Mai, T., Xu, X., Zhang, P., Li, M., & Liu, Y. (2018). NetworkAI: An intelligent network architecture for self-learning control strategies in software defined networks. *IEEE Internet of Things Journal*, 5(6), 4319–4327.

[26] Khurshid, A., Zou, X., Zhou, W., Caesar, M., & Godfrey, P. B. (2013). VeriFlow: Verifying network-wide invariants in real time. 2013.

[27] Kazemian, P., Chang, M., Zeng, H., Varghese, G., Mckeown, N., & Whyte, S. (2013). Real time network policy checking using header space analysis. In *Proceedings of the 10th USENIX Conference on Networked Systems Design and Implementation* (pp. 99–111).

[28] Hinrichs, T., Gude, N., Mitchell, J., Shenker, S., & Berkeley, U. C. (2009). Expressing and enforcing flow-based network security policies. (pp. 1–20).

[29] Ristenpart, T., & Tromer, E. (2009). Hey, you, get off of my cloud: Exploring information leakage in third-party compute clouds. In *Proceedings of the 16th ACM Conference on Computer and communications security* (pp. 199–212).

[30] Yao, G., Bi, J., & Xiao, P. (2011). Source address validation solution with OpenFlow / NOX Architecture. In *Proceedings of the 2011 19th IEEE International Conference on Network Protocols* (pp. 7–12).

[31] Li, D., Hong, X., & Bowman, J. (2011). Evaluation of security vulnerabilities by using ProtoGENI as a launchpad. In *Proceedings of the 2011 IEEE Global Telecommunications Conference - GLOBECOM 2011*.

[32] Fonseca, P., Bennesby, R., Mota, E., & Passito, A. (2012). A replication component for resilient OpenFlow-based networking. In *Proceedings of the 2012 IEEE Network Operations and Management Symposium* (pp. 933–939).

[33] Kreutz, D., Ramos, F. M. V., & Verissimo, P. (2013). Towards secure and dependable software-defined networks. In *Proceedings of the Second ACM SIGCOMM Workshop on Hot Topics in Software Defined Networking* (pp. 55–60).

[34] Kl, R., & Smith, P. (2013). OpenFlow: A security analysis. In *2013 21st IEEE International Conference on Network Protocols (ICNP)*.

[35] Shin, S., & Gu, G. (2013). Attacking software-defined networks: A first feasibility study. In *Proceedings of the Second ACM SIGCOMM Workshop on Hot Topics in Software Defined Networking* (pp. 165–166).

[36] Christianson, B., Malcolm, J, Stajano, F., Anderson, J., & Bonneau, J. (2013). *Security Protocols XXI*. In 21st International Workshop.

[37] Wen, X., Chen, Y., Hu, C., & Wang, Y. (2013). Towards a secure controller platform for OpenFlow applications. In *Proceedings of the Second ACM SIGCOMM Workshop on Hot Topics in Software Defined Networking* (pp. 171–172).

[38] Shin, S., Yegneswaran, V., Porras, P., & Gu, G. (2013). AVANT-GUARD: Scalable and vigilant switch flow management in software-defined networks. In *Proceedings of the 2013 ACM SIGSAC Conference on Computer & Communications Security* (pp. 413–424).

[39] Hutchison, D. (2013). Cyberspace Safety and Security. In *5th International Symposium*.

[40] Smeliansky, R. L. (2014). SDN for network security. In *Proceedings of the 2014 International Science and Technology Conference (Modern Networking Technologies) (MoNeTeC)*.

[41] Yi, A., Crowcroft, J., Tarkoma, S., & Flinck, H. (2014). Software defined networking for security enhancement in wireless mobile networks. *Computer Networks*, 66, 94–101.

[42] Li, H., Member, S., & Li, P. (2014). Byzantine-resilient secure software-defined networks with multiple controllers in cloud. In *Proceedings of the IEEE Transactions on Cloud Computing* (pp. 1–12).

[43] Scott-Hayward, S., Kane, C., & Sezer, S. (2014). OperationCheckpoint: SDN Application Control. In *Proceedings of the 2014 IEEE 22nd International Conference on Network Protocols*.

[44] Abou El Houda, Z., Hafid, A. S., & Khoukhi, L. (2019). Cochain-SC: An intra-and inter-domain DDoS mitigation scheme based on blockchain using SDN and smart contract. *IEEE Access*, 7, 98893–98907.

[45] Hu, H., Ahn, G., Han, W., & Zhao, Z. (2014). Towards a reliable SDN firewall. In *Proceedings of the 2014 Open Networking Summit Research Track (ONS 2014)*.

[46] Ge, M., Hong, J. B., Yusuf, S. E., & Kim, D. S. (2018). Proactive defense mechanisms for the software-defined Internet of Things with non-patchable vulnerabilities. *Future Generation Computer Systems*, 78, 568–582.

[47] Porras, P., Cheung, S., Fong, M., Skinner, K., & Yegneswaran, V. (2015). Securing the software-defined network control layer. In *Proceedings of the Network and Distributed System Security Symposium*.

[48] Hu, D., Hong, P., & Chen, Y. (2017, December). FADM: DDoS flooding attack detection and mitigation system in software-defined networking. In *Proceedings of the GLOBECOM 2017-2017 IEEE Global Communications Conference* (pp. 1–7). IEEE.

[49] Chin, T., & Mountrouidou, X. (2015). Selective packet inspection to detect DoS flooding using software defined networking (SDN). In *Proceedings of the 2015 IEEE 35th International Conference on Distributed Computing Systems Workshops*.

[50] Fung, C. (2015). FlowRanger: A request prioritizing algorithm for controller DoS attacks in software defined networks. In *Proceedings of the 2015 IEEE International Conference on Communications (ICC)* (pp. 5254–5259).

[51] Chi, P. W., Kuo, C. T., Guo, J. W., & Lei, C. L. (2015). How to detect a compromised SDN switch. In *Proceedings of the 2015 1st IEEE Conference on Network Softwarization (NetSoft)*.

[52] Kamisinski, A., & Fung, C. (2015). FlowMon: detecting malicious switches in software-defined networks. In *Proceedings of the ACM CCS Workshop on Automated Decision Making for Active Cyber Defense 2015* (pp. 39–45).

[53] Menezes, D., Mattos, F., Carlos, O., & Bandeira, M. (2016). AuthFlow: Authentication and access control mechanism for software defined networking. *Annals of Telecommunications*, 71, 607–615.

[54] Ambrosin, M., Member, S., Conti, M., Member, S., De Gaspari, F., & Poovendran, R. (2016). LineSwitch: Tackling control plane saturation attacks in software-defined networking. In *Proceedings of the IEEE/ACM Transactions on Networking* (pp. 1–14).

[55] Aleroud, A. (2016). Identifying DoS attacks on software defined networks: A relation context approach. In *Proceedings of the 2016 IEEE/IFIP Network Operations and Management Symposium* (pp. 853–857).

[56] Pang, C., Jiang, Y., & Li, Q. (2016). FADE: Detecting forwarding anomaly in software-defined networks. In *Proceedings of the 2016 IEEE International Conference on Communications (ICC)*.

[57] Zhang, M., Bi, J., Bai, J., Dong, Z., Li, Y., & Li, Z. (2017). FTGuard: A priority-aware strategy against the flow table overflow attack in SDN ∗. In *Proceedings of the SIGCOMM Posters and Demos* (pp. 141–143).

[58] Xu, T., Gao, D., Dong, P., Foh, C. H., & Zhang, H. (2017). Mitigating the table-overflow attack in software-defined networking. In *Proceedings of the IEEE Transactions on Network and Service Management* (pp. 1–12).

[59] Zhou, Y., Chen, K., Zhang, J., Leng, J., & Tang, Y. (2018). Exploiting the vulnerability of flow table overflow in software-defined network: attack model, evaluation, and defense. *Security and Communication Networks* (pp. 1–15).

[60] Zhang, Q. Y., Wang, X. W., Huang, M., Li, K. Q., & Das, S. K. (2018). Software defined networking meets information centric networking: A survey. *IEEE Access*, 6, 39547–39563.

[61] Li, F., Cao, J., Wang, X., & Sun, Y. (2017). A QoS guaranteed technique for cloud applications based on software defined networking. *IEEE Access*, 5, 21229–21241.

[62] Son, J., & Buyya, R. (2018). A taxonomy of software-defined networking (SDN)-enabled cloud computing. *ACM Computing Surveys (CSUR)*, 51(3), 1–36.

[63] Au, N. N. H., & Pham, V. H. (2019). Toward a trust-based authentication framework of Northbound interface in software defined networking. In *Proceedings of the Industrial Networks and Intelligent Systems* (p. 269). Springer Nature.

[64] Hussain, K., Hussain, S. J., Jhanjhi, N. Z., & Humayun, M. (2019). SYN flood attack detection based on Bayes Estimator (SFADBE) for MANET. In *Proceedings of the 2019 International Conference on Computer and Information Sciences (ICCIS)* (pp. 1–4). IEEE.

3 Clustering in Wireless Sensor Networks Using Adaptive Neuro-Fuzzy Inference Logic

Seema Gaba, Radhika Gupta, Sahil Verma, Kavita, and Imran Taj

CONTENTS

3.1   INTRODUCTION

Numerous applications of self-organized networks that operate without any centralized support have tempted many researchers to study this area. Wireless sensor networks are the most well-known types of self-organized networks being used in many applications. These networks are called ad-hoc networks. The main provision of creating these networks is to readily provide services [1] to end users. The absence of infrastructure or central coordination works to the advantage of mobile ad-hoc networks, or MANETs, due to reduced cost, complexity, and, most importantly, time required for network deployment. Apart from these reasons, MANETs are also suitable for emergency situations like natural or human-induced disasters, military conflicts, emergency medical situations, and business solutions, etc. MANET introduces its branches based on the network and application requirements. The main reason these networks succeed is on-the-go setup. Apart from this capability, a sensor network's success depends on the availability of information, type of information, quality of information, duration of applicability [2], and a couple of other details. It is very important to send the collected information to the base station to fulfil the network's deployment purpose. These type of networks have their use in many fields, including the medical and military fields. So, it is important that data transfer is robust and fault tolerant due to constrained resources, limited bandwidth, and a few other limitations; it becomes important that for efficient data

DOI: 10.1201/9780367808228-3

transfer, the network should be fault tolerant and have a longer life. Network lifetime ultimately relies upon the energy utilization of individual nodes in the network. In one of the studies, authors have discussed challenges that will likely present when industrialization reaches its peak. Security and availability are mentioned among those challenges, which comprise future research of WSNs. Because the sensors depend on battery power, that power must be used [3] wisely to prevent unnecessary use. In the previous study, clustering is given as a reliable solution to conserve the node's energy, hence improving network lifetime. The clustering concept reduces [4] the amount of information received at base station from each and every node and provides managed load balancing by organizing the network in a connected hierarchy. Clustering may follow one hop [5] or multi-hop communication. Clustered networks often suffer from the hot spot problem, where cluster heads near the base station die much earlier due to the load of data forwarding. Such deaths of data-forwarding nodes result in network coverage loop holes. However, clusters can be made energy aware in different environments by different methods. The point of concern is the selection of cluster heads. Low energy adaptive clustering hierarchy (LEACH) is one of the prominent routing protocols for sensor networks; it uses the concept of clustering. In this protocol, [6] heads are chosen on the basis of probability value. The chances are that a node with low energy [7] or at a far distance from the sink node may get selected as head. The heads are required to be more capable to collect information [8] from their members, filter it, and send it to sink. Therefore, a lot of new work has been applied to overcome the shortcomings of LEACH. One of the studies combines LEACH, mobile sink, and other points to utilize LEACH and improve CH selection. This paper proposes a fuzzy logic-based technique to provide more refined selection of cluster heads. The fuzzy-inference system is a decision-making system that draws outputs from fuzzy inputs, i.e., uncertain parameters. The logic works on IF-THEN rules and deals with reasoning on the high level using information acquired from users. Fuzzy logic provides an adjustable support to address the aspects of dynamic networks [9] with major uncertainties, which leads to failures and overloads. Adaptive neuro fuzzy is a type of artificial neural network in which neuro-adaptive learning techniques provide a fuzzy modeling method to learn about particular data sets. In basic terms, neuro fuzzy is the tuning of one of the fuzzy inference systems, named sugeno, by using training data. It combines the learning power of an artificial neural network and the explicit knowledge representation of a fuzzy inference system. Key features of this system include use of input-output patterns to adjust the inference rule system. The information is collected and then the inference system is trained to recognize the patterns and behave accordingly. To summarize, we need a training data set for the appliance of neuro-fuzzy logic. In the field of wireless sensor networks [10], adaptive neuro-fuzzy logic has been used for various purposes. The tuned sugeno fuzzy logic can be used in processes like data aggregation, sensing, and remote environmental monitoring in sensor network. For example, [11] proposed an ANFIS based algorithm to monitor temperature gathering in a sensor network. The main advantage is that using ANFIS instead of Mamdani fuzzy type can give better results in terms of accuracy. ANFIS technique is also used to optimize network parameters like calculation of inner packet gap, measurement of latency period, and others. A few studies say that ANFIS gives good outcomes when input parameters do not exceed five. More inputs can cause computational complexes, hence causing delay in calculations. Considering this outcome, in this paper, fewer than five parameters are taken.

The successful outcomes of ANFIS can be based on the robustness of results it provides. ANFIS has highly generalized capabilities of machine-learning techniques or neural networks. ANFIS can take crisp input values, represented in the form of membership functions and fuzzy rules, and also generate crisp output of fuzzy rules for reasoning purposes.

The remaining paper is organized as a review of the existing work done for wireless sensor networks using fuzzy logic, followed by the proposed technique and simulation results of different networks. The last section of the paper covers the conclusion and future work in this context.

3.2   RELATED WORK

Dhananjay Bisen et al. (2018) worked on improving the performance of the AODV routing protocol by reducing the broadcast of unnecessary hello messages. They used a Mamdani fuzzy inference system and adaptive neuro-fuzzy inference system (ANFIS) to calculate the resultant optimal interval of the hello transfer. Energy level and mobility speed of the nodes are used as input for the inference system. The proposed technique is helpful, but it cannot be applied for sensor network because AODV is not a routing protocol for sensors. It can be checked with routing protocols like LEACH and others to be used for sensor networks.

K. P. Vijayakumar et al. (2018) authors have used the fuzzy approach to protect the network against a jamming attack. The fuzzy approach uses two network metrics as input, namely, the packet delivery ratio and received signal strength to detect jamming. To detect jamming, authors say the fuzzy approach gives better results compared to true detection ratio.

Thanga Aruna Muthupandian et al. (2017) presented a survey for techniques for selection of forwarding node (cluster head) in a sensor network. Along with the cluster head paper, the authors also tell about the selection of the next node to be selected for data transmission to send it across nodes to the base station. Authors have presented a comparative study of the fuzzy logic and adaptive neuro-fuzzy logic based on network parameters like the packet delivery ratio, distance, traffic, energy, and others.

Mohammad Shokouhifar, Ali Jalali (2017) has proposed a new energy-optimized routing algorithm for wireless sensor networks called LEACH-FS [12]. To provide uniformity in making an efficient cluster, the algorithm uses a fuzzy c-means method along with bee colony to adjust the fuzzy rules. The fuzzy system selects the appropriate cluster head for the network. The inputs given to fuzzy inference are distance from sink, residual energy, and distance from the cluster centroid. The artificial bee colony algorithm is used to automate the tuning of rules. This automatic tuning eliminates the manual loading of rules.

S. A. Sahaaya et al. (2017) proposed an enhanced zone-stable, fuzzy-logic based cluster head election algorithm (ZSEP-E) [13] for a wireless sensor network. Fuzzy logic inputs are remaining energy, density, and the distance from sink node of the network. A zone partition algorithm forms three zones, with two zones having homogeneous sensor nodes with same capabilities. Nodes are not location aware and not made to be mobile. Also, the initial energy of nodes is categorized as low, intermediate, and advanced, indicating the static nature of selecting heads as nodes; nodes with high energy are the most probable candidates for selection as the cluster head. Fuzzy rule-based optimization is given as a future scope of the work.

B. Baranidharan et al. (2016) have given a distributed load-balancing technique for clustered networks. Authors say load balancing plays an important role so that the network can serve for longer. The fuzzy approach is used for selecting the cluster head. The input parameters for a Mamdani inference system are residual energy, highest number of neighbours, and distance from base station. The cluster size is kept unequal so that the cluster head near the base station can serve for longer, keeping the cluster small. Taking this work forward, the neuro-fuzzy concept is experimented with in this paper to check real-time parameters after every data transmission.

Julie (2016) has proposed an artificial intelligence technique: adaptive neuro fuzzy inference system is used, which forms neuro-fuzzy energy aware clustering scheme (NFEACS) [14]. Using the neural network property of ANFIS, signal strength and energy are trained. Therefore, based on errors found in testing data against training data of these two factors, fuzzy-logic cluster heads are selected. The sink location is centralized here. More refined results are possible if one or more parameters, like node degree, are taken for training set data.

M. Selvi, R. Logambigai et al. (2016) Paper presents a fuzzy temporal logic for clustering. The process differs in that two relay nodes, called a cluster head [15] and super cluster head, are formed. The super cluster head performs routing across clusters, and the cluster head transfers data within clusters. The fuzzy-rule combinations have proven ineffective in decision making. However, the metrics like flow control and congestion control are not considered in this work, which leads to improved quality of service.

Zahra Beiranvand et al. (2013) presents an energy-efficient algorithm, improved LEACH (I-LEACH) [16], which considers location and a number of neighbour nodes, along with energy and distance from sink node, to select appropriate cluster head. However, once deployed, the network is kept static, i.e., the mobility factor of the ad-hoc sensor network is not taken for this work.

Hakan Bagci et al. (2013) has proposed a fuzzy approach to solve the complications of hot spot in network. The algorithm [17] forms clusters of different size, and the clusters near the sink node are smaller in size, reducing intra-cluster work. The mobility factor of sensors is not considered in this work, which may produce more efficient results.

Krasimira Kapitanova et al. (2012) used fuzzy rule-based inference system for robust event detection in sensor networks. However, the authors say that it is difficult to tune the exponentially growing size of rule base. Instead of using crisp values, it is preferred to use fuzzy values for handling uncertainty. The issue of managing a larger rule base can be managed if we have enough data sets that can be trained with existing rules. This can be done applying artificial intelligence with fuzzy logic. This is what paper has tried to implement using adaptive neuro-fuzzy logic.

Safdar AbbasKhan et al. (2012) authors, for detection of faults in the sensor networks, implement another application of the fuzzy inference system. Each node in the network has its own fuzzy model, which is based on a rule set. The rule set gets input as the sensor measurements of neighbouring nodes. Based on this, the node's actual measurement is outputted. Based on the difference in actual measurement and fuzzy output measurement, a node is declared faulty. Participation of faulty node in data transmission can lead to security breach and network shutdown. Therefore, using fuzzy logic, authors have provided a better approach.

Hasung-Pin Chang et al. (2012) presents a hybrid communication protocol named cross-layer energy efficient protocol (CEEP) [18] to minimize information exchange back and forth in the network by the cross-layer optimization concept. The hybrid routing is applied by combining both reactive and proactive schemes, thereby implementing proactive routing within zones and reactive between zones. Authors have only two regions in the network, which can create additional overhead if the network is bigger. In that case, even the intra-zone routing can cause delay to reach the sink, which can be in another zone.

Sudip Misra et al. (2010) has given a simple least-time energy-efficient protocol with one-level data collection (LEO) [19]. The protocol is of a proactive nature, with a modification that every node only contains information about its neighbours. One-level aggregation is done at the node, which is in the closest vicinity of the sink node. Protocol provides a route that takes less time to transfer data from nodes with higher energy and less distance to sink. However, factors like mobility and security are not considered for this work.

Hoda Taheri et al. (2012) paper proposes an energy-aware distributed dynamic clustering protocol for the clustering in sensor networks. Authors have chosen tentative cluster heads based on their remaining energy, and then for a set of nodes, fuzzy logic is applied to check the node fitness for electing as head. The paper presents use of if-then sugeno fuzzy rules to have final selection of cluster head. If considering other network parameters, this work can provide more reliable results.

Toleen Jaradat et al. (2013) have proposed a fuzzy-based cross-layer routing for sensor networks. The fuzzy inference system is inputted three parameters, remaining battery life, link quality, and transmission power for nodes. The combination of if-then rules of these parameters gives the output as a node, which will be next relay node. Every time, fuzzy logic will select the next node, which will communicate the data.

Nimisha Ghosh et al. (2017) an energy-efficient routing is proposed [20–32] for wireless homogeneous sensor network, with the use of a mobile collector. The genetic algorithms, which include particle swarm optimization (PSO) and ant colony optimization (ACO), are applied for the results optimization. The main theme follows the selection of cluster head from the clusters in network using fuzzy logic. The mobile collector gathers information from these heads and delivers to sink node. The LEACH routing protocol is improved in a manner where, rather than selecting the cluster head in each round, it is selected on demand. This on-demand selection will reduce the overhead caused during each round in the earlier practices. The linguistic variables for fuzzy logic used here are node degree, packet drop probability, and node centrality. On the basis of these variables, a value's chance for a node to become a cluster head is calculated. PSO is used to optimize the membership function for the better range outputs (chance).

3.3 PROPOSED WORK

The proposed technique aims to optimize the network efficiency by improving network performance parameters and packet-delivery ratio, hence optimizing network energy consumption. The neuro-fuzzy logic is used to help in selecting appropriate cluster heads. The algorithm of the proposed work is as follow:

Step 1. Read the total number of nodes and initial energy.

Step 2. Initialize cluster and cluster head to 1.

Step 3. Check for changed/updated energy values of nodes.

Step 4. For all nodes in network form the clusters. If energy constraint satisfies, then check this condition $c(i)<=(P/(1-P*mod(r,(1/P))))$ and increase candidate head count. Here $c(i)$ is random value for each node between 0 and 1 and P and r are the head probability and iteration, respectively.

Step 5. Store cluster head id and energy and calculate distance from base station.

Step 6. Calculate packet drop for each candidate node by using $1-(10*log10$ (distance.^error_rate)) here, distance is distance between candidate head and other members of cluster.

Step 7. For all selected candidate nodes for head input energy, distance from sink and packet drop value to fuzzy model.

Step 8. Check for the testing data (cluster head parameters fed to neuro-fuzzy model) with training data. If obligatory output, train the parameters by adjusting energy and distance.

Step 9. Calculate the minimum distance route to perform routing.

Step 10. Check the required results.

3.4 SIMULATION RESULTS

The algorithm was implemented in Matrix Laboratory (MATLAB®) using MATLAB scripting language. Table 3.1 represents the simulation parameters taken. The network was set up randomly deployed in an area of M*M dimensions with n number of nodes and a sink node. Initially, each node has 2J of energy. A minimum distance of 30 m is set for communication between nodes.

The experiment was run for varying nodes at different iterations. Figure 3.1 represents the energy-consumption graph for the case 1, 50 nodes run for 25 iterations. It can be seen from the graph that at the 25th iteration, only 32% of

TABLE 3.1
Simulation Metrics

Simulation Metrics	
Parameters	**Values**
Area	200 m * 200 m
Energy Level	2J
Mobility Model	Random way point
Number of Nodes	Varied
Threshold distance	30 m
Number of iterations	Varied

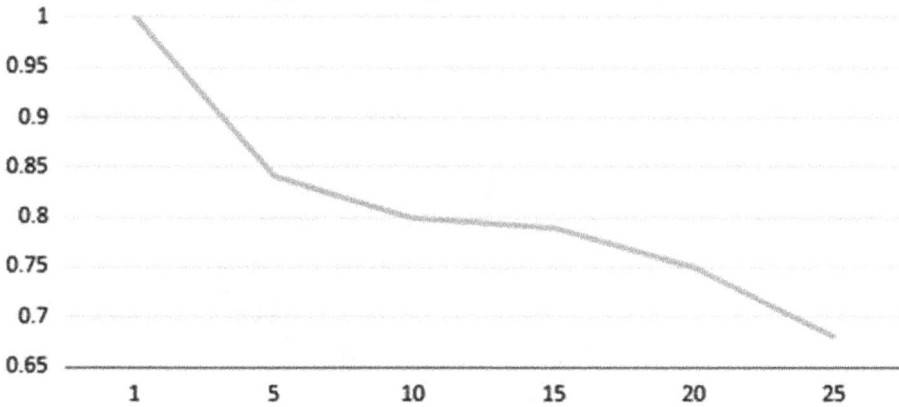

FIGURE 3.1   Energy consumption for 25 rounds of 50 nodes.

energy is consumed and 68% energy is still preserved. The packet-delivery ratio is one of the important network-performance parameters also noted for 50 nodes on 25 rounds. Table 3.2 represents the packet-delivery ratio values obtained for case 1.

Figure 3.2 represents the graph for packet-delivery ratio of 50 nodes for case 1. The average packet-delivery ratio obtained in this case is 97.58%.

The simulation was run again for case 2 having 50 nodes for 50 iterations. Figure 3.3 represents the energy consumption graph for this case. It is seen from the graph that on 50th iteration 75% of energy is consumed and 25% is still left. Packet delivery ratio values obtained for this case are shown in Table 3.3.

Figure 3.4 represents the graph for packet-delivery ratio of 50 nodes for case 2. The average packet-delivery ratio obtained on the 75% utilization of energy is 99.59%.

TABLE 3.2
Packet Delivery Ratio for Case 1

Rounds	Values of packet delivery ratio
1	1
2	1
3	1
4	1
5	1
6	1
7	.996
8	.991
9	.985
10	.98
11	.976
12	.97
13	.965
14	.96
15	.957
16	.953
17	.95
18	.948
19	.945
20	.94

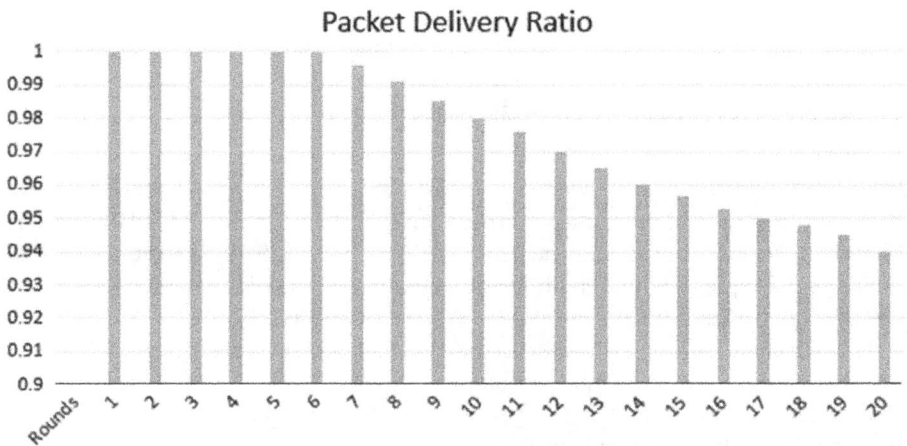

FIGURE 3.2   Packet delivery ratio for case 1.

Energy consumption for 50 rounds

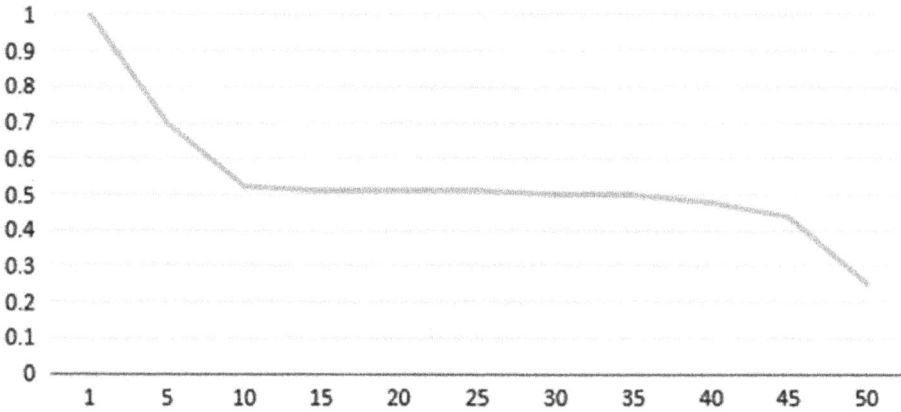

FIGURE 3.3 Energy consumption for 50 rounds of 50 nodes.

TABLE 3.3
Packet Delivery Ratio for Case

Rounds	Values of packet delivery ratio
1	1
2	1
3	1
4	1
5	1
6	1
7	.998
8	.998
9	.997
10	.997
11	.996
12	.995
13	.995
14	.9945
15	.994
16	.994
17	.993
18	.991
19	.991
20	.99

Packet Delivery Ratio

FIGURE 3.4   Packet delivery ratio for case 2.

Energy consumption for 25 rounds

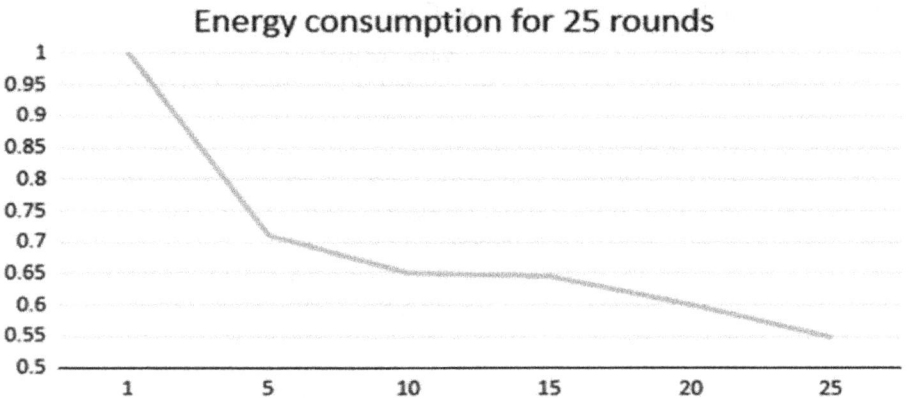

FIGURE 3.5   Energy consumption for 25 rounds of 100 nodes.

Case 3 had the 100 nodes, and simulation was run for 25 rounds. Figure 3.5 represents the energy-consumption graph for this case. It is seen from the graph that on the 25th iteration, 45% of energy is consumed and 55% is still preserved.

Table 3.4 represents the values of packet-delivery ratio obtained for the case 3.

Figure 3.6 represents the graph for packet-delivery ratio of 100 nodes for case 3. The average packet-delivery ratio obtained on the 45% utilization of energy is 99.61%. Next was testing 100 nodes for 50 iterations.

Figure 3.7 represents the energy consumption graph for case 4. The graph depicts almost full consumption of energy at 50th iteration for 100 nodes.

In Figure 3.8, graph for packet delivery ratio of 100 nodes for case 4. The average packet-delivery ratio obtained on the 99.8% utilization of energy is 99.54%.

TABLE 3.4

Packet Delivery Ratio for Case 3

Rounds	Values of packet delivery ratio
1	1
2	1
3	1
4	1
5	1
6	1
7	.998
8	.998
9	.9965
10	.996
11	.995
12	.995
13	.995
14	.9945
15	.994
16	.993
17	.993
18	.991
19	.99
20	.99

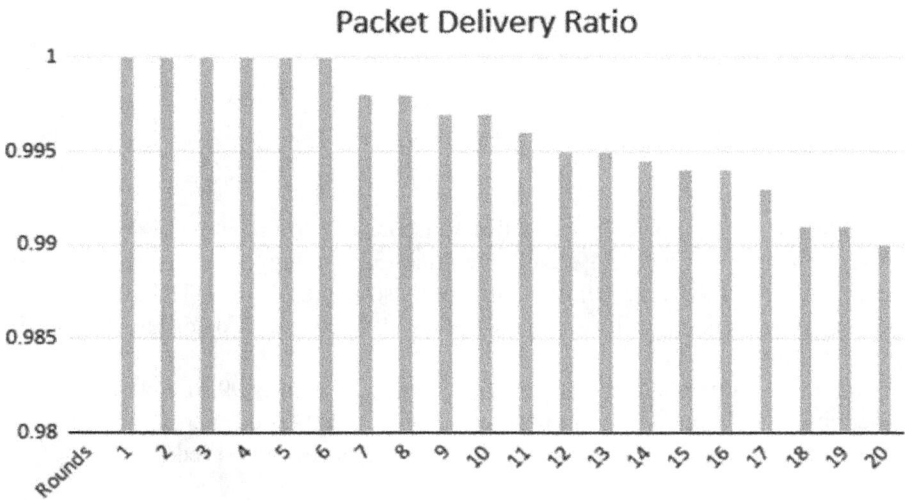

FIGURE 3.6   Packet delivery ratio for case 3.

Energy consumption for 50 rounds

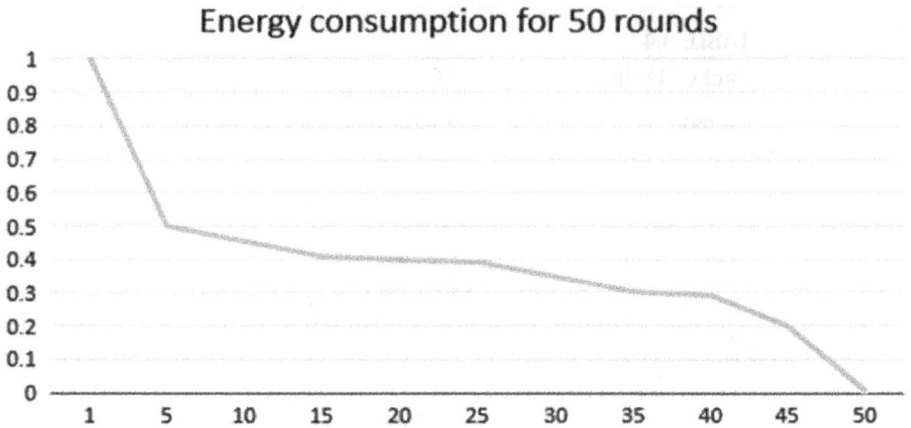

FIGURE 3.7   Energy consumption for 50 rounds of 100 nodes.

Packet Delivery Ratio

FIGURE 3.8   Packet delivery ratio for case 4.

Table 3.5: The graphs shows that at full consumption of energy, the obtained packet-delivery ratio is 99.54% and when only 32% energy is consumed, then the results for packet-delivery ratio obtained are also good, 97.58%. The performance metrics values, energy utilization, and packet-delivery ratio for different cases is given in Tables 3.6 and 3.7.

Maximum value obtained for the packet-delivery ratio is 99.61%, which is obtained at 45% of energy consumption. On 45% utilization of energy, good results are obtained; therefore, more than 50% of energy is still preserved.

TABLE 3.5
Packet Delivery Ratio for Case 4

Rounds	Values of packet delivery ratio
1	1
2	1
3	1
4	1
5	.998
6	.997
7	.997
8	.9965
9	.9965
10	.996
11	.996
12	.995
13	.995
14	.9945
15	.994
16	.992
17	.992
18	.991
19	.99
20	.99

TABLE 3.6
Performance Parameters for 50 Nodes

Number of nodes = 50

Number of Iterations	Energy Consumption (%)	Packet Delivery Ratio (%)
25	32	97.58
50	75	99.59

3.5 CONCLUSION AND FUTURE SCOPE

Wireless sensor networks are deployed in thousands of numbers; hence, clustering is considered as one of the best ways for preserving energy of the nodes during communication. The proposed work aims to improve the selection of cluster heads in the network. Use of adaptive neuro-fuzzy logic includes the artificial behaviour in network in a way that data is tested against an already trained set to fit into the

TABLE 3.7

Performance Parameters for 100 Nodes

Number of nodes = 100

Number of Iterations	Energy Consumption (%)	Packet Delivery Ratio (%)
25	45	99.61
50	99.8	99.54

requirement. The fuzzy inference system works on rule sets, and combining it with artificial intelligence (ANFIS) can provide better-optimized results, which are received in our experiment as per considered metrics. The simulation results have shown good results for both the parameters considered here. However, the network's efficiency is not based on only these two parameters. As we have mentioned, other authors have used this technique with other parameters; like delay, throughput can also contribute in testing the network performance. Considering more metrics may result in more optimized and improved results. However, with multiple network metrics, it may become difficult to have a larger rule base if the fuzzy approach is used, but techniques for load optimization of rule base can be tested. Also, mobility is an important factor, which can become the part of neuro-fuzzy logic for better results.

REFERENCES

[1] H. Hasani and S. Babaie, "Selfish node detection in ad hoc networks based on fuzzy logic," *Neural Computing and Applications* 3, 1–12, 2018.
[2] A. Abuarqoub, M. Hammoudeh, B. Adebisi, S. Jabbar, A. Bounceur, and H. Al-Bashar, "Dynamic clustering and management of mobile wireless sensor networks," *Computer Networks*, 117, 62–75, 2017.
[3] Mian Ahmad Jan, Bin Dong, Syed Rooh Ullah Jan, Zhiyuan Tan, Sahil Verma, and Kavita, *A Comprehensive Survey on Machine Learning-Based Big Data Analytics for IoT-Enabled Smart Healthcare System MONET*, Springer, 2021.
[4] X. Liu and Xuxun, "A Survey on clustering routing protocols in wireless sensor networks," *Sensors*, 12(8), 11113–11153, 2012.
[5] K. Akkaya and M. Younis, "A survey on routing protocols for wireless sensor networks," *Ad Hoc Networks*, 3(3), 325–349, 2005.
[6] E. G. Julie, S. Tamilselvi, and Y. H. Robinson, "Performance analysis of energy efficient virtual back bone path based cluster routing protocol for WSN," *Wireless Personal Communications* 91(3), 1171–1189, 2016.
[7] H. Yaser, S. Ali, N. Z. Jhanjhi, M. Humayun, A. Nayyar, and M. Masud, "Role of fuzzy approach towards fault detection for distributed components," *CMC-Computers Materials & Continua* 67, (2), 1979–1996, 2021.
[8] Y. G. Ha, H. Kim, and Y. C. Byun, "Energy-efficient fire monitoring over cluster-based wireless sensor networks," *International Journal of Distributed Sensor Networks*, 2012, 2012.

[9] M. Humayun, N. Z. Jhanjhi, and M. Z. Alamri, "Smart secure and energy efficient scheme for E-Health applications using IoT: A review," *International Journal of Computer Science and Network Security* 20(4), 55–74.

[10] I. Batra, S. Verma, Kavita, U. Ghosh, J. J. P. C. Rodrigues, G. N. Nguyen, A.S.M. Sanwar Hosen, and V. Mariappan, "Hybrid logical security framework for privacy preservation in the green internet of things," *MDPI-Sustainability* 12(14), 5542, 2020.

[11] I. J. Su, C. C. Tsai, and W. T. Sung, "Area temperature system monitoring and computing based on adaptive fuzzy logic in wireless sensor networks," *Applied Soft Computing* 12(5), 1532–1541, 2012.

[12] S. Hafsa, H. Ashraf, H. Javed, M. Humayun, N. Z. Jhanjhi, and M. A. AlZain, "Energy optimised security against wormhole attack in IoT-based wireless sensor networks," *CMC-Computers Materials & Continua* 68(2), 1966–1980, 2021.

[13] S. A. Sahaaya, A. Mary, and J. B. Gnanadurai, "Enhanced zone stable election protocol based on fuzzy logic for cluster head election in wireless sensor networks," *International Journal of Fuzzy Systems*, 19(3), 799–812, 2017.

[14] S. Soobia, et al., "Analyzing the performance and efficiency of it-compliant audit module using clustering methods," in *Proceedings of the Industrial Internet of Things and Cyber-Physical Systems: Transforming the Conventional to Digital. IGI Global*, 351–376, 2020.

[15] M. Selvi, R. Logambigai, S. Ganapathy, L. S. Ramesh, H. K. Nehemiah, and K. Arputharaj, "Fuzzy temporal approach for energy efficient routing in WSN," in *Proceedings of the International Conference on Informatics and Analytics*, 1–5, 2016.

[16] Z. Noor, L. T. Jung, F. Alsaade, and T. Alghamdi, "Wireless sensor network (WSN): Routing security, reliability and energy efficiency," *Journal of Applied Sciences* 12(6), 593–597, 2012.

[17] H. Mamoona, N. Z. Jhanjhi, M. Alruwaili, S. S. Amalathas, V. Balasubramanian, and B. Selvaraj, "Privacy protection and energy optimization for 5G-aided industrial internet of things," *IEEE Access* 8, 183665–183677, 2020.

[18] H. Chang, "A hybrid intelligent protocol in sink-oriented wireless sensor networks," in *Proceedings of the 2012 International Conference on Information Security and Intelligent Control*, 57–60, 2012.

[19] S. Maryam, H. Ashraf, A. Ullah, M. Masud, M. Azeem, N. Z. Jhanjhi, and M. Humayun, "Robust cluster-based routing protocol for IoT-assisted smart devices in WSN," *CMC-Computers Materials & Continua* 67(3), 3505–3521, 2021.

[20] N. Ghosh, I. Banerjee, and R. S. Sherratt, "On-demand fuzzy clustering and ant-colony optimisation based mobile data collection in wireless sensor network," *Wireless Networks* 1–17, 2017.

[21] H. S. Jawad, M. Irfan, N. Z. Jhanjhi, K. Hussain, and H. Mamoona, "Performance enhancement in wireless body area networks with secure communication," *Wireless Personal Communications* 116(1), 1–22, 2021.

[22] P. Rani, Kavita, S. Verma, and N. G. Nguyena, "Mitigation of black hole and gray hole attack using swarm inspired algorithm with artificial neural network," *IEEE Access* 30, 2020.

[23] B. Vijayalakshmi, K. Ramar, N. Z. Jhanjhi, S. Verma, M. Kaliappan, K. Vijayalakshmi, Kavita, U. Ghosh, and S. Vimal, *An Attention Based Deep Learning Model For Traffic Flow Prediction Using Spatio Temporal Features Towards Sustainable Smart City*, IJCS,Wiley, 2021.

[24] S. Kar, S. Das, and P. K. Ghosh, "Applications of neuro fuzzy systems: A brief review and future outline," *Applied Soft Computing*, 15, 243–259, 2014.

[25] A. Tatar, A. Barati-Harooni, A. Naja_-Marghmaleki, B. Norouzi-Farimani, and A. H. Mohammadi, "Predictive model based on an_s for estimation of thermal conductivity of carbon dioxide," *Journal of Molecular Liquids* 224, 1266–1274, 2016.

[26] S. Verma, S. Kaur, D. B. Rawat, C. Xi, L. T. Alex, and N. Z. Jhanjhi. "Intelligent framework using IoT-based WSNs for wildfire detection," *IEEE Access* 9, 48185–48196, 2021.

[27] M. M. Afsar and M.-H. Tayarani. "Clustering in sensor networks: A literature survey," *Journal of Network and Computer Applications* 46,198–226, 2014.

[28] R. Madan, et al., "Cross-layer design for lifetime maximization in interference-limited wireless sensor networks," *Wireless Communications* 5(11), 3142–3152, 2006.

[29] J.-H. Chang and L. Tassiulas, "Maximum lifetime routing in wireless sensor networks," *IEEE/ACM Transactions on Networking (TON)* 12(4), 609–619, 2004.

[30] R. Madan and S. Lall, "Distributed algorithms for maximum lifetime routing in wireless sensor networks," *Wireless Communications* 5(8), 2185–2193, 2006.

[31] A. Giridhar and P. R. Kumar, *"Maximizing the functional lifetime of sensor networks,"* in Proceedings of the 4th International Symposium on Information Processing in Sensor Networks. IEEE Press, 2005.

[32] S. Mottaghi and M. R. Zahabi, "Optimizing LEACH clustering algorithm with mobile sink and rendezvous nodes," *AEU-International Journal of Electronics and Communications* 69(2), 507–514, 2015.

4 Security in Big Data

Mehwish Malik, Hina Umbrin, Nuzhat Akram,
Khalid Hussain Usmani, and NZ Jhanjhi

CONTENTS

DOI: 10.1201/9780367808228-4

4.1 BIG DATA

To begin with, it is a common belief that big data describes huge data sets that need novelties in analytical methods to fully utilize them and create new and innovative kinds of value. The vastness of big data is not because of absolute magnitude or size; instead, it's about the appropriate scale of studies and analysis. Researchers have defined it in various ways since it's a ubiquitous term exploited in various parts of academia and industries. Sagiroglu [1] associated big data to volumes of data sets, saying big data is an expression for large-scale data sets having huge, more diverse, as well as complex, structures with the complications of analysing, storing, and visualizing data for additional processing and results. Another definition proposed by Van Dijck [2] stated big data as social action transformation to online measured data, which consequently allow predictive analysis and real-time tracking. They encourage aspirations to build more reliable and accurate predictions to resolve complex and intricate problems, ranging from changes in climate to terrorist activities Kitchin [3]. Furthermore, big data embody administrative challenge regarding extensive information accumulation by corporate ventures, as well as state agencies. Bekker [4] described big data as the data characterized by informational features, such as statistical correctness and the nature of event logs, etc., and that urges such technical necessities as parallel processing, distributed storage, and uncomplicated solution scalability. The author further described each feature comprehensively, arguing that traditional data was suspectable to change at any time; for instance, bank accounts, product counts at a warehouse, etc., whereas big data depicts a log of each record denoting certain events, for example, web page view, purchase activity in a store, sensor value w.r.t time period, social media comments, etc. This nature of big data allows data of events to not change. Though big data is considered statistically correct, it may contain errors or omissions, which is why it is not considered a good choice for tasks requiring absolute accuracy. Another property of big data highlighted by the author is its technical requirements due to its volume; it requires parallel processing, and high-storage capacity needs a special storage approach.

Andrea [5] identified core themes linked to big data, which are information, method, technology, and its impact. They have quoted various explanations of the term big data and checked if they possess the aforementioned themes or not. Information can be termed as one of the fundamental reasons behind the existence of big data, as its generation and availability serve as fuel to big data. Information in big data gave rise to a new term called datafication, which aims to organize a digital form of analog signals to produce insights that couldn't have been inferred otherwise. Technology is frequently associated with big data, which allows its exploitation such as Hadoop [6,7]. The capability to efficiently store an extensive amount of data on relatively smaller machines is a fundamental element of technology application on big data. Hence, it's justified to say that technology is the core equipment to work with big data. Extensive analysis of quantitative data and a necessity to grab the value from an individual's behavior need processing techniques and methods; thus, methods transform big data into an asset. The exploitation of big data analytics not only allows to manage data efficiently and properly but also help incorporation of such data for decision-making processes. Awareness of these methods, along with technologies, their strength and weaknesses, and cultural tendencies, spread to facilitate informed and intelligent decision making, as required by big data. The degree of impact it is imposing on our society is often portrayed through success stories and anecdotes of technology and methods of implementation. These stories combined, with novel principles and procedural developments, lead to a valuable contribution toward knowledge creation on a subject. If the pervasive quality of information availability and productions results in tons of applications spanning several scientific fields, then it also has an adverse impact on society as well. Major concerns arising due to the evolution of big data are: privacy issues, issues regarding information accessibility, etc.; henceforth, impact is an integral theme of big data. Authors [5] then grouped the definitions into a few groups, the first being the group focusing entirely on enlisting characteristics of big data. In this group, Laney [8] presented a framework that expresses the increase in volume, variety, and velocity of data in all three dimensions; this framework was later named as three V's of big data. This model was later extended to value [9], veracity [10], variability [11], viscosity [12], virality [12], and validity [13,14], described in detail below. Another group defined big data in terms of technical requirements behind processing huge amounts of data. The rest of the definitions associated with big data to crossing certain thresholds, such as when data surpass the processing capacity of traditional database systems, then it can be termed as big data. A conclusive definition was proposed [5] in light of previous definitions; it states that big data is a representation of informational resources characterized by high velocity, volume, and variation to necessitate specific analytical processes and technology to transform it into value. To cap it all, big data is simply a transformation where data is processed into information, which is processed into knowledge, which turns into wisdom and value, as demonstrated in Figure 4.1 (Big Data Transformation).

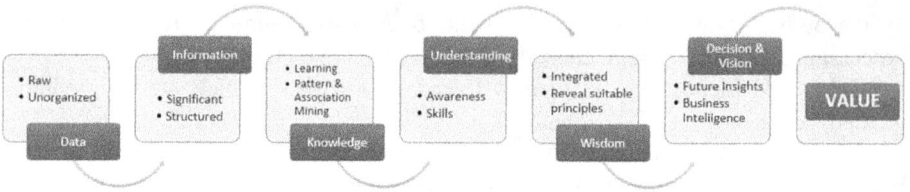

FIGURE 4.1   Big Data Transformation.

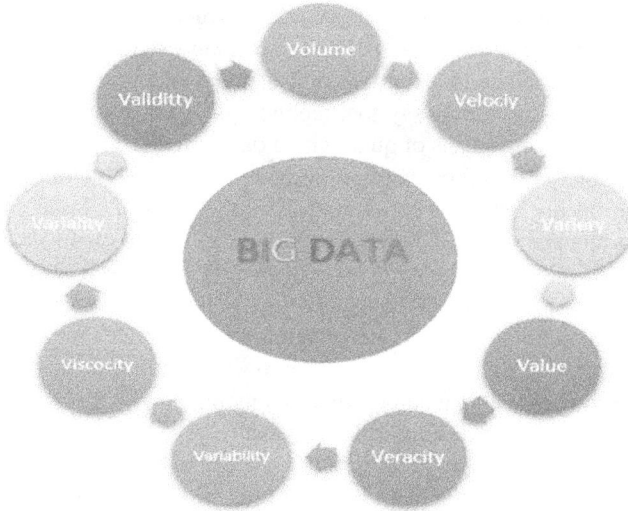

FIGURE 4.2   9 V's of big data.

To sum it all, big data shapes a phenomenon based on technology, culture, and scholar knowledge that stands on the interaction of technology, mythology, and analysis that provokes wide dystopian and utopian rhetoric Boyd [15].

Big data characteristics were initially described by Laney [8] as part of a data framework, not directly associated to big data but was made part of the big data model later and was then extended by several authors. These characteristics are explained below and also illustrated in Figure 4.2.

4.1.1   Volume – The Size of Data

The sheer volume is principle trademark that makes the data 'big.' Volume is surely the base, if we consider big data as a pyramid. It states the enormous amount of data being produced every second from cell phones, social media, M-2-M sensors, credit cards, videos, photographs and so forth. The present quantity of data can be pretty staggering. Consider a few examples:

- In an annual report prepared by Thomas Reuter in 2010, he shared his estimation of world data to be more than 800 exabyte and sill growing further. In 2012, a hardware company EMC claimed world data to be 99 exabyte and predicted its growth rate to be 50% every year.
- Up to 500 videos are getting uploaded on YouTube each minute as per the statistic computed till May 2019.
- More than a trillion photos were captured in 2018, and that number is expected to rise by 7% in current year.
- In 2018, worldwide mobile traffic added up to 19.01 exabytes every month. In 2022, this data traffic is likely to reach nearly 77.5 exabytes every month globally, at a consolidated growth rate of about 46% annually. The measurements that used to be in gigabytes are now measured in zetta or even yottabyte due to the exponential growth of data.
- According to quite recent figures, Facebook generates approximately 4 petabyte of data every day. This biggest social networking platform has around 2.41 billion active users each month, according to the facts retrieved by quarter in 2019. It also has 400 users signing up for its service every minute; in that one minute, not only are accounts created, but also, 293 thousand statuses are created, 510 thousand comments are made, 136 thousand pictures are uploaded, and around 4 million posts get likes.

It's still ambiguous about how much data is generated every year, but the amount of information processed is no doubt huge. This immense volume of data needs different and distinct processing technologies, rather than relying on traditional processing and storage capabilities. At the end of the day, big data is too massive to be processed on any ordinary laptop processor; thus, we cannot analyse and store data using conventional database technologies. Now, distributed systems are utilized, and they store chunks of data at different places and can be brought back into single units by software. Collection and analysation of such vast amounts of data is a big challenge.

4.1.2 Variety – Different Forms of Data

Variety can be stated as different forms of data that can be used. Today's data is certainly different from data in past years, which was mostly structured data; such data includes bank statement information, including amount, date, account title, contact information, etc. Traditional data fits perfectly fine in any relational database. Big data handles diverse types of data, which can be divided into three main categories: structured, semi-structured, and unstructured data. Let's briefly discuss each one.

- *Structured Data* – Such data is well formatted and organized data, which has defined length for big data. Its examples are dates, numbers, combination of numbers, and words named as strings, etc. Such data is often stored in a relational database.

- *Semi-structured Data* – It lies between structure and unstructured data. It's a kind of structural data, which does not follow data model's formal structure linked to data tables or relational databases. Nonetheless, it contains markers or tags to detach semantic elements and impose records and fields hierarchies within data. Example include XML files and JSON; NoSQL also is considered a semi-structured type.
- *Unstructured Data* – Currently, most of the world's data (around 80%) reside in an unstructured category of data. It includes pictures, videos, updates on social media, such as tweets, statuses, posts, etc., voice recordings, CCTV footages; in addition to these, it also contains log files, machine data, click data, sensor data, etc. Unstructured data augment structured data, where things like audio files, web pages, MRI images, twitter feeds, and web logs are places. As evident, it contains everything that can be stored and captured, but it is not based on a meta model (collection of rules to surround an idea or concept) that precisely defines it. Unstructured data can better be defined when compared to structured data. Structured data can be thought of as data that is well defined under established rules; for instance, names are depicted as text, numbers will be used for money with minimum two decimal points, and there's a specific pattern followed by dates. Whereas in unstructured data, no rules are followed; for instance, a tweet, voice recording, or picture all are different but represent thoughts and ideas built on human understanding.

The variety of data types requires special algorithms with diverse processing capabilities. Organizing data to extract meaningful information is no ordinary task, particularly when its changing at a rapid pace. Big data technology's innovation and novelty have somehow enabled harvesting, utilization, and storage of all type of data.

4.1.3 VELOCITY – SPEED OF DATA GENERATION

Velocity denotes the speed at which an enormous amount of data is getting generated, as well as created, collected, refreshed, and analysed. It's the incoming data frequency that needs processing. Twitter messages, SMS messages, swipes of credit cards, and Facebook status updates sent over specific telecom carriers at each minute, are generated at high velocity. One of the popular streaming applications that manages data velocity is a web service by Amazon called Kinesis. Few examples to apprehend the idea behind velocity are listed here.

- Google processes more than 70 thousand queries every second, making up to 4 million searches going on per minute, 240 million in each hour, and more than 5.76 billion searches each day. It reflects the phenomenon of change in our lives due to the internet.
- Facebook claims the per day incoming data rate to be more than 500 terabytes, for which it has a dedicated data warehouse at Prineville, Oregon. Some 240 billion pictures or more are stored by Facebook, in addition to

350 million new pictures being uploaded each day by users. To save these photos, storage of 7 petabyte gear each month is deployed by the data centre team of Facebook. Though it looks remarkable that Facebook stores data of more than 300 petabytes, a significant factor that should be accounted is the pace of creation of new data, aka its velocity. The speed with which data is increasing imposes the need for data analysis, but it also demands the data access and transmission rate to be prompt to enable real-time access to instant messaging, verification of credit cards, and access to websites. It's often argued that quick flow of information, as real time as possible, is a basic requirement of companies and that velocity can turn out to be more critical and important than volume since it provides greater competitive benefit. Its said that it's better to have limited data access in real time than low access rate to huge data. Availability of data at the right time helps suitable decision making in businesses; after a certain time period, that data may not be as significant as it was before.

4.1.4 VALUE – DATA'S WORTH

By value, it means the worth of extracted data. An endless amount of data can turn into useless stock unless it has some value. The value characteristics of big data sit at the topmost position in a pyramid of big data, which signifies the ability to change data's tsunami into business.

The trade-off between cost and benefit for analysing and collecting data helps understand if it's monetary to reap the data, and it's the most crucial phase of embarking on true initiative of big data. Considerable value can be created in big data, which enables customers to understand well, targeting them according to that understanding, improving business performance, and optimizing methods. It's important to understand any strategy before embarking on it, along with its potential and challenges. The question arises whether gathered insights will help in creating a new line of product, cost-cutting measure, or cross-selling opportunity, or will it aid in discovery of crucial underlying effects that may produce a cure to a certain disease? If a company exploits big data correctly, after substantial investment on resources and time, then it can possess the ability to understand its customers, and, along with that, can monetize enormous information. That can allow the company to give offers that meet their customer needs at right time.

4.1.5 VERACITY – DATA UNCERTAINTY

Veracity is trustworthiness, reliability, or quality of data. It states how accurate the data is. Take for an example, Twitter posts with abbreviations, hashtags, typos, etc., and the accuracy and reliability of such content. Another related example is GPS data usage, when users visit urban areas, GPS will sense off course. Tall buildings will bounce back satellite signals; so, in such situations, another source, such as road data, would be required to be fused with location data to provide correct information. This feature of big data is considered unfortunate since the increase in any of the rest of the V's causes a drop in veracity; this may be similar to volatility and validity of big data (defined below). Gleaning loads of data is useless if it's not accurate.

Data-veracity knowledge enables better understanding of analysis, and risks associated with it, which eventually assists better decision making in business. Veracity ensures the accuracy and cleanliness of data, which keeps your system away from bad data accumulation by using various processes. Data with high veracity contains several records that are valuable for analysis and can contribute to complete results in a meaningful way, whereas data with low veracity contains a high proportion of meaningless data, often called noise. Medical trial or experiment data sets have high veracity.

4.1.6   VARIABILITY – DATA INCONSISTENCY

Variability in big data refers to few different concepts. First, it often means data inconsistencies, which are required to be found by outlier or anomaly detection techniques to do useful analytics. Secondly, variability in big data is because of dimensional data multitude, which is the result of numerous disparate data sources and types. Finally, it may state inconsistency in speed of loading big data to traditional databases. Variability is not the same as variety; say, for an example, four different coffee blends are offered by coffee shop, but if a customer gets the same blend each day, but finds it tastes different every day, then it's variability.

4.1.7   VISUALIZATION – DATA REPRESENTATION

Another important feature of big data is visualization challenge. Visualization discusses how data is represented through graphs or charts, etc. Representing huge quantities of complex data using charts and graphs, etc., is comparatively more effective than reports and spreadsheets packed with formulas and numbers to convey meaningful information. Existing visualization tools of big data are facing some technical difficulties because of in-memory technology limitations, poor functionality, scalability, and response time. To plot billions of data points, traditional graphs are not so useful; therefore, a need to represent data arises using various methods of data representation, such as tree maps, data clustering, parallel coordinates, cone trees, circular network diagrams, etc. Integrating this approach with a variable's multitude occurs from big data's velocity, variety, and complicated relationships among them.

4.1.8   VOLATILITY – HOW LONG TO STORE DATA

How long can your data be kept until it is considered historic, irrelevant, or useless? How old should your data be to be considered useful? Before the arrival of big data, companies used to keep data indefinitely – a small number of terabytes may not produce excessive storage expenses; it may also be possible to store it in live databases without inducing performance issues. Due to volume and velocity of big data, volatility should be carefully considered as well. Volatility states how long the data should be stored and still be valid. Real-time analysis on data requires the determination of the point at which data is not pertinent to ongoing analysis. A need to establish new rules for data availability and currency, as well as rapid information retrieval, emerges when the situation asks for it. These rules should be tied to business processes and needs with emphasis on complexity and cost of retrieval and storage process of big data.

4.1.9   VALIDITY – DATA USE

Like veracity, it also emphasizes the accuracy and correction of data for its expected use. As per Forbes, data scientists spend approximately 60% of their time on data cleaning before they even start analysis. Big data analytics are as essential as the underlying data; hence, it is desirable to adopt decent governance approaches to guarantee consistent quality, metadata, and definitions of data.

There are several other V's proposed by authors [16,17] such as viscosity, virality, vulnerability, vocabulary, and so forth. The above nine characteristics are further grouped into a few categories by authors. Veracity and variety come under data collection; velocity and volume come under the category of data processing; validity, variability, and volatility come under the category of data integrity. Alternatively, visualization goes in data visualization, and value is observed in data worth. These categories help in determining the gist behind each feature of big data.

4.2   DATA SOURCES OF BIG DATA

Big data consists of various types of cumulative data. Following are the key sources for big data: public, private, and community data; data wear out and self-quantified. Public data is the data seized by government institutions and local communities and transferred by business and management applications, such as transportation, energy use, and health care that required individual privacy under specific conditions. Private data are data contained by private companies, non-profit firms, and personal information that cannot be revealed by public sources; examples include consumer transactions, identification tags used by institutional supply chains, internet browsing, and usage of mobile phone. Data wear out is the type where data is collected from limited or zero-value data collection partners and attached with other resources for the purpose of creating new data. Another cause of such type of data is because of our information-seeking requirements for behaviours that can be utilized to conclude people needs or objectives. Community data is an unstructured data that includes consumer reviews on goods and voting review, for instance, twitter feeds, Facebook comments, and so on. Self-quantification is exposed from action and behaviours; for example, a digital watch customized to display exercise and movement. Such a self-quantification type of data helps to create a bridge between psychology and behaviours [18].

4.3   ARCHITECTURE

Big data design is the all-encompassing framework used to ingest and transform huge amounts of data with the goal to analyse the data for business decisions. The architecture is a blueprint for a solution of big data, grounded on the organization's business needs. Underneath headings, the layers involved in the architecture are precisely discussed.

4.3.1   DATA SOURCE

The organizations produce a huge proportion of data nearly once a day, and it is developing exponentially. The data-source layer integrates the data coming in from various sources, at various speeds, and in different associations.

4.3.2   INGESTION

Big data ingestion includes interfacing with different information sources, extracting the information, and distinguishing the transformed information. Following are the parameters for data ingestions: data velocity, data size, data frequency, and data format [19]. The data has been passed through from the following layers.

1. *Identification* – Arranged into different observed data structure; the unstructured data is entrusted with default structures.
2. *Filtration* – The data applicable for the end is constantly separated based on the (MDM) repository. MDM stands for Management.
3. *Validation* – After filtration, data is analysed in contrast of MDM metadata.
4. *Reduction of Noise* – By removing the noise and minimizing the inconsistencies, data is cleaned.
5. *Transformation* – Data is separated or combined based on its types, substance, and the necessities of the organization.
6. *Compression* – The size of the data is decreased without impacting its significance for the required method. It should be seen that weight does not impact the result assessment.
7. *Integration* – The refined dataset is united with the software utility layer, e.g, Hadoop.

4.3.3   STORAGE LAYER

This is the key part for any big-data based system. It impacts the versatility, data structures and programming, and computational models of the system [20].

4.3.4   STAGING

A staging is a middle of storage area utilized for data handling during the extraction, transformation, and loading (ETL) process.

4.3.5 Data Pipeline

The veracity of big data requires a quality of data coming in and out of the big-data processing pipeline. A big-data pipeline is divided into two parts: micro and macro pipeline. Micro pipeline work is based on level steps to make sub forms. A micro pipeline includes a granular data-handling step. Macro pipelines operate on workflow level and control each level workflow properly.

4.3.6 Data and Workflow Management

Big data architecture is built on large-scale distributed groups of data with scalable capacity. If a big data environment is built on a cloud, there is a need to spend time to establish a strong agreement with a cloud provider.

4.3.7 Data Access

The data structure exceptionally relies upon how applications or clients need to retrieve the information. Data-retrieval patterns should be known because some types of information can be redundantly recovered by enormous numbers of clients or applications (Figure 4.3). The big data architecture is shown in Figure 4.3 for the reader better understanding.

4.4 BIG DATA CHALLENGES

Big data is more like a torrential slide that's rapidly advancing down the mountain, getting faster and greater along the way, with a majority of organizations scrambling to keep pace with it. This is just like a skier requiring the fundamental equipment, such as helmet, gloves, etc., to survive an avalanche. Likewise, organizations need to be prepared for the avalanche, which, in this case, is big data; they need to be aware of essential tools while the avalanche is gaining steam. Having comparatively more opportunities to collect data from far more data sources is what makes data 'big'. Think about every one of the billions of gadgets that are currently internet-abled, cell phones and IoTs [21] sensors being just two cases. Though big

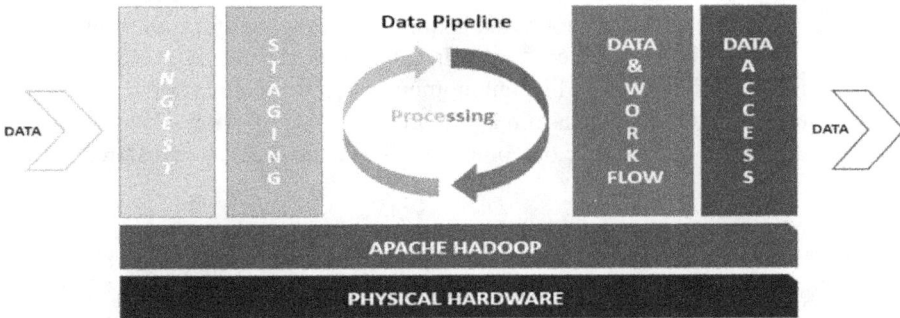

FIGURE 4.3 Big data architecture.

data offers several attractive opportunities, still professionals and researchers face many challenges while exploring large data sets and when they extract knowledge and value from such information. The complexity of these huge data sets poses plentiful difficulties at different levels, such as data storage, data capturing, sharing, searching, management, analysis, and visualization of data. Now imagine about all security-related problems that can arise. The idea of big data appeared from an unbelievable growth in the figure of IP-equipped devices. On a lighter note, big data is simply a term for entirely accessible data or information in a specified region that a corporation gathers with the objective of discovering trends or unseen patterns within it [22]. These, after getting discovered by analytical tools, can be utilized to yield an improved result not far off (more revenue, higher satisfaction level of customers, fast delivery of services, and so on). The other side of the coin states that the architecture for big-data storage reflects a new target of its security concerns for malware and illegal activities. The question is, should something happen to such an integral business asset, the outcomes could be disastrous for the association that accumulated it. Sadly, several big-data tools are open source, which are not often designed to keep security as a major function, prompting yet increased security issues. Often, the deluge of distributed streams and information surpasses our ability to harness.

Prior to going to battle, every general needs to study his enemies: the size of their army, type of weapons, battle count with results, their tactics, etc., in order to craft right strategy for battle and be prepared for the fight. Similarly, every decision-maker needs to comprehend what they are managing. Here, some major challenges attributed to big data are discussed below (Figure 4.4). Big data has a number of challenges and opportunities, as shown below.

4.5   BIG DATA ANALYTICS CHALLENGES IN BIG DATA

Big data carries huge transformative potentials and opportunities for different areas; but, at the same time, it additionally exhibits extraordinary challenges to binding such huge expanding volumes of data. Cutting-edge investigation is needed to interpret the associations among the features and examined data. For instance, data analysis empowers an association to fetch useful insights and keep track of the patterns that might influence business negatively or positively. There are various additional data-driven applications that require real-time analysis, such as social networks like Facebook, Twitter, LinkedIn, etc., area of finance, navigation, intelligent transportation systems like autonomous cars, astronomy, medicine, and so forth. Therefore, effective data-mining approaches, along with advanced algorithms, are required for accurate outcomes to observe the changes in different fields

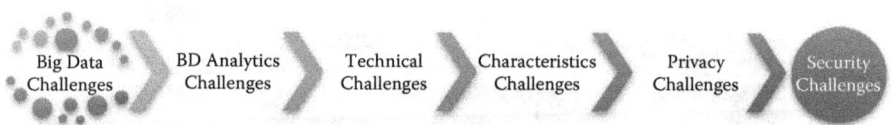

Big Data Challenges → BD Analytics Challenges → Technical Challenges → Characteristics Challenges → Privacy Challenges → Security Challenges

FIGURE 4.4   Major big data challenges.

and predict future perceptions [23]. Nonetheless, big data analysis is yet very challenging for numerous reasons: intricate nature of it, such as the Seven V's, the necessity for performance and scalability to scrutinize such enormous heterogenous data sets with realistic responsiveness [24–26]. These days, there are several analytical techniques including statistical analysis, data mining, visualization, and machine learning. Numerous studies deal with this area either by enhancing already proposed techniques or by proposing new or a tested ensemble of approaches or algorithms. Hence, big data accelerated the development of hardware, software, and system architectures. However, despite everything, there's a need of analytical progression to confront challenges of big data. Another issue is the assurance of timely response when the data volume is huge. In spite of the fact that scientists are trying to improve quality of data, as well as make analytical algorithms further robust (resistant to problems related to data), big-data analytics is not flawless or ideal. It's simply not yet possible to tackle some of the issues associated with data's reliability. Poor analysis leads to erroneous conclusions and correlations [23]. In the below section, major challenges encountered in big-data analytics are explored.

a. *Data Management Landscape's Uncertainty* – Due to the fact that big data is expanding day in and day out, new technologies and companies are being established every day. Finding out the best technology without initiating new problems or risk is a great challenge for companies [27].

b. *Big Data Capacity Gap* – Although big-data analytics is a rising field, the number of experts of this field is very limited. The reason behind insufficient experts is because it's a complex field and very few people understand the intricate and complex nature of this area. Hence, the talent gap presence in this industry is another big challenge for big data.

c. *Data attainment into big data platforms* – Companies are facing the challenge of handing this huge amount of rapidly increasing data. The variety and scale of today's data can overwhelm any data expert, which is what make it more crucial to make data easily available for managers.

d. *Synchronization among data sources* – Diversity of data sets increase the need to consolidate them into some analytical platform. In case this is overlooked, it can generate gaps, leading to incorrect insights and message [28].

e. *Important insights generation through big data analytics* – It is significant that organizations obtain legitimate insights from big-data analytics, and it's also equally important to right department approaches to this data and access it. Thus, it's a major challenge to bridge this gap in productive fashion.

f. *Biased markers in Big Data Analytics* – The big-data algorithms are frequently grounded on explicit markers associated with the analysed element. And because of this, analytics can be misrepresented. When someone cracks what markers impact the results, they can modify the analysed element to satisfy the prerequisites set by the markers. As stated, that big-data analytics depend on specific markers of the studied item. If the individual appending the marker is biased toward the matter, it will influence the outcome. Hence, biased markers result in biased analysis.

4.6   TECHNICAL CHALLENGES IN BIG DATA

It tends to be anything but difficult to get lost in the assortment of big data technologies currently available in the market. Do you require Apache Spark, or would the pace of MapReduce be sufficient? Is it desirable to use HBase or Cassandra for data storage? Discovering answers to these questions can be complicated. And it's easy to pick inadequately, on the off chance that you are searching the ocean of technological prospects without a clarity of what you require [23].

These technical difficulties are further divided to subheadings defined below.

a. *Fault Tolerance* – The arrival of new technologies such as big data and cloud computing increase the expectations that every time a failure occurs, the harm done should be within tolerable threshold as opposed to starting the entire task from scratch. However, fault tolerance is incredibly hard computing, involving complex algorithms. It's basically impractical to devise completely foolproof and 100% reliable machine or software. Thus, the fundamental assignment is to decrease the likelihood of failure/error to an acceptable stage. It is unfortunate that the more we struggle to decrease this probability, the higher the expense. There's a strategy to overcome this problem. It divides the entire calculation being performed into smaller tasks and allocates these tasks to various nodes for the sake of computation. One of the nodes is given the responsibility to check proper working of these nodes. In case something happens, that specific task is restarted. However, there can be a scenario where it's not possible to divide the entire computation into independent tasks. The tasks can be recursive where the input of previous task depends on the computation of next one. So, it can be cumbersome to restart the whole computational task [23].

b. *Scalability* – The technology behind the processor has changed recently. The speed of clock has slowed down to a great extent, whereas the number of cores has increased in a processor. Earlier data-processing frameworks needed to stress over parallelism across nodes in a specific cluster, but recently the concern has moved to intra-node parallelism. The techniques used in past for parallel-processing across nodes aren't efficient enough to handle parallelism within a single node. It's because considerably more hardware resources are jointly used across a core in a separate node. The issue of scalability in big data has led to cloud/distributed computing, which presently aggregates many dissimilar workloads with fluctuating performance objectives into huge clusters. This necessitates a higher level of resource sharing, which is costly and furthermore carries with it many challenges, such as how to execute and run different jobs so we can cost-effectively achieve the goal of every workload. It additionally requires managing the failures in an efficient way that arises more often than if operating on big clusters [29].

c. *Data Quality* – Huge amount of data collection and storage emanates a substantial cost. Better results can be attained if more data is fed to predictive analysis and decision-making processes in any business. Business leaders always want more data storage; on the other hand, IT leaders consider technical aspects before data storage. Big data essentially focuses on quality of data storage, rather than having irrelevant bulk data, in order to draw better outcomes and conclusions. This further prompts different queries, like how it very well maybe guaranteed what data is relevant, how much of the data would be sufficient for the process of decision-making, and even if the data stored is correct or how not to reach inference from it, etc.

d. *Data Heterogeneity* – Unstructured data embodies nearly every sort of data being constructed, for instance, recorded gatherings, PDF documents handling, social media interactions, emails, fax transfers, and many more. Structured data is constantly being organized in extremely automated and manageable ways. It depicts proper integrations with a database; however, unstructured data is totally crude and disorderly, So, working with such data is obviously exorbitant and cumbersome. Conversion of such unstructured data to organized data is not feasible, either. Digging through this data is unwieldy as compared to structured data that be easily managed [24].

4.7 CHARACTERISTICS-ORIENTED CHALLENGES OF BIG DATA

Every property of big data possesses a challenge, few of them of major concern are discussed below.

4.7.1 DATA VOLUME

The very first issue associated with this property is related to storage. The amount of space increases with the increase in data volume, so there is a need to store it efficiently. Apart from the fact that gigantic volumes of data should be retrieved at a quick pace to fetch output from them, there are other areas that need attention, as well such as storage costs like cloud storage versus in-house storage of data, bandwidth, networking etc. [30]. With the expansion in data volume, the worth of data records reflects decreases in proportion to richness, age, type and quality [31]. The advent of social-networking websites has led to a generation of data of the order of peta/terabytes each day. Aforementioned data volumes are hard to tackle using existing conventional databases [31].

4.7.2 DATA VELOCITY

More and more data is being generated on a regular basis by computer systems, and both analytical and operational speeds as well as the amount of consumers of said data are growing. Individuals need the entirety of data, and they need it at the earliest possible time, prompting to what is popular as high-velocity data. Data of high velocity denotes millions of columns of data every second. Classic outdated database systems are not fit enough for executing analytics on such data, which is continuously in motion. Unfortunately, the state-of-art technology can't manage data that restrict the collection of data, such as data generated by machines and human activities like Twitter feeds, log files, website clicks, and mortar, etc. [31].

4.7.3   DATA VARIETY

Big data comes in numerous fashions, namely messages, images, and updates in social-networking sites. Global-positioning systems signals come from smartphones, sensors, and many more. A considerable number of these big data sources are almost new or somewhat as old as social-networking websites similar to Facebook and Twitter launched in 2004 and 2006, respectively. Cellular phones and other such devices can be put in the same class. The ubiquitous nature of these devices makes traditional databases unsuitable for these data. Wide range of these data is unwieldy, noisy, and unstructured, which entails thorough techniques for data-based decision-making. Hence, improved algorithms for analysing such data is a critical issue [32].

4.7.4   DATA VALUE

Companies store data to gain useful insights from it and utilize it for business-intelligence analytics. This process of storing creates a gap between IT professionals and business leaders. As mentioned before, business experts are found concerned over data value for their business and how can the data be profitable for them. They believe that more data generates more insights, whereas IT experts also deal with technical details when storing and processing big data [31].

4.8   PRIVACY CHALLENGES

Although people appreciate the ease brought by big data, they experience numerous inconveniences as well. If big data is not very safe for user data during the time of use, it will legitimately undermine the protection of users and data security. As per many protection contents, it tends to be subdivided into categories such as anonymous protection, anonymous identifiers, and privacy protection. A person's privacy is a very important and sensitive issue that holds technical, conceptual, and legal significance [33]. The private data of a person when joined with external huge datasets leads to new facts inference about that individual; for example, vendors posting relevant ads after observing users' spending habits, such as on specific designer clothes, styles, or locations, etc. It is possible that the individual may not want the data owner or any other person to know these secret facts. Currently, numerous organizations believe that the identifiers will be covered up, after processing the information anonymously. However, security assurance can't be viably accomplished through the use of anonymous protection. Users information is gathered and used so as to increase value to organization's business. It's made possible by forming insights in users' private lives, about which they are not aware. Another significant consequence emerging is called social stratification where an aware and literate person takes benefits from big data's analytical ability [34] and the underprivileged or naïve users will be identified easily, hence treated awfully. Many countries lack regulations and rules for management of user information in the current era and don't have decent supervision systems. All this coupled with user's unawareness of self-protection has instigated many issues because of leakage of information [35].

4.9 SECURITY CHALLENGES

Security and privacy are huge challenges in the era of big data, and this concern is increasing with the growth of data each second [36]. A key reason behind this is that information is now easily and widely accessible. Professionals from various areas, such as medicine, business, and government, are sharing data on a large-scale with each other. Yet, the developed tools and technologies in use until today are not competent enough to handle such enormous amount of data and are not efficient enough to give satisfactory security [37]. The technologies are deficient to provide enough maintenance features for privacy and security due to the absence of a fundamental understanding about ways of providing security to such an enormous bulk of data and because necessary training is not offered concerning how to equip these large-scale data sets with security and privacy [37]. Big data security, together with privacy maintenance in regard to big data, lacks suitable policies that guarantee compliance with recent approaches in security and privacy. Current technologies are continuously encountering accidental and intentional breaches because of having weak privacy and security-maintenance abilities. Therefore, it's necessary to update and reassess present methods to prevent continuous data leakages. It has also been realized that IT dompanies spend a minuscule financial resource on big data protection. Nearly 10% of an organization's IT-related budget ought to be spent on security, but less than 9% on average is spent, consequently making it difficult for these companies to ensure protection of data [37].

4.10 CURRENT SECURITY CHALLENGES IN BIG DATA

Before digging deep into security concerns of big data, let's first define big data security to get better understanding of each aspect associated to this subject.

4.10.1 Big Data Security – A Definition

Big data security is a joint term for every tool and measure used to protect data, as well as analytics processes from theft, attacks, or other similar malicious practices that can in any way harm or adversely influence them. Similar to other types of cyber security, variants of big data are also concerned with offenses that get initiated either from offline or online domains. For organizations working on cloud, this challenge is multifaceted. These dangers include online information theft, DDoS attacks, server crashes, and ransomware. The matter can be much more problematic when organizations store confidential or sensitive information, such as credit card numbers, customer personal information, or even just contact particulars. Furthermore, these attacks on big data storage organizations could originate serious financial aftermaths like fines, losses, sanctions, or litigation expenses, etc.

Big data is nothing novel to bigger organizations. However, its popularity is not restricted to large organizations; instead, it equally holds the same status among small and medium-size firms. The reason behind it is its simplicity in terms of management; along with this, it comes with the benefit of reduced cost [38]. Cloud storage has encouraged the use of data collection and data mining. Be that as it may, this big data integration with cloud-based storage has instigated challenges to security and privacy threats. The cause behind such breaches is probably because that the applications are designed to accumulate a certain quantity of data, but they are unable to define a mechanism of handling such a gigantic volume of data that the previously mentioned data sets have. Similarly, these security technologies are not efficient enough to cope with dynamic data and are only suitable to control the static data. While the common public reaps the benefits of big data, they also confront numerous inconveniences. If the data of users is not protected well, then it results in security and privacy concerns. A few security case studies are discussed in the upcoming section, and after that, challenges are discussed.

4.10.2 CASE STUDIES OF SECURITY BREACHES DEPICTING THEIR IMPACT ON ORGANIZATIONS

According to Cyber Edge [39], more than 60% of 763 security professionals participating in a study disclosed effective cyber-attacks on their medium to large corporations. A report in 2015 by Data Breach Investigation corresponded about 80K security occurrences, in addition to 2.122K affirmed data breaches [40]. A couple of years ago, a solitary data breach affected around 1–10 million records of a target company but these days, even a single data breach can result in a compromise of around 0.2 billion records, triggering multimillion-dollar loss and harm to brand names, alongside confronting governing penalties. Consider Ashley Madison [41], a site for extramarital affairs, which endured a data breach in 2015 and suffered data leakages of 25 gigabytes. This not only caused harm to the reputation of this social website but also brought about disturbances in clients' lives, with reports of two suicide cases. One more instance is of Target [42], a discount retailer in the United States, through a data rift occurring during Christmas season in 2013. The breach occurred at POS (point of sale) system where hackers stole information of about 0.04 billion debit and credit cards. In addition to that, they also gained unauthorized to data of around 70 million consumers, which included user names, email addresses, phone numbers, etc. A report by Naked Security stated that this data break had cost about $290 million, of which insurance charges covered $90 million. In spite of that, the company is still facing lawsuits by the shareholders and inquiries by state attorney and the Federal Trade Commission, which could further cost them over $30 million, collectively a loss of $300 million.

Take another example of New Jersey based Horizon Blue Cross Shield (HSC) [43]. The company encountered two similar cases, in 2013 and 2016, respectively. In January 2008, the company modified its corporate policy because of the theft of an employee's laptop. The policy stated to implement encryption on all devices such as mobiles, desktop systems, and laptops. Even then, in 2013's act of data breach, many laptops were found unencrypted by the investigating officials. Again, another data leakage occurred and the company realized that laptops were protected by passwords, but had no encryption implemented on them. This breach exposed customer names, birthdates, SSNs, insurance identification numbers, and clinical information. This made the company to pay $1.1 million for inferior security solutions. Comparable case is of Yahoo [44–46]. A data breach happened around 2013–2014, which was reported in September 2016 when Yahoo was in negotiations with Verizon. They reported that they suffered the biggest breach ever in the history, where about 3 billion user accounts were sacrificed. Around the same time, 117 million users lost information on LinkedIn [44,47], a networking website of professionals, when it was hacked and a massive number of email addresses and passwords were purchased by the black market. Another biggest data breach is of Facebook [48], where nearly 50 million user profiles were collected for Cambridge Analytica with intention to influence voters' choice in the US election. The the analytics firm collaborated with current US President Donald Trump's team. The Cambridge Analytica team, along with Steve Bannon (personal advisor) used private information of users in an unauthorized manner to construct a system that would profile each voter to target them with custom political ads. In a very recent case in March 2019, Toyota issued an official statement on their website confirming a data breach, which hypothetically affected around 3.1 million users. Several other cases of data breaches that affected large population of users have occurred; only a few of them are stated above, but even from these examples, it is evident that big data can cause serious security problems. It is by these bitter experiences that IT and business specialists are learning to improve security and protect big data.

4.11 MAJOR SECURITY ISSUES OF BIG DATA

Here's a list of few of the security challenges [49] that may occur while using big data.

4.11.1 DISTRIBUTED FRAMEWORKS SECURITY

The majority of big data applications distribute massive processing jobs over several systems for rapid analysis. A eminent example of it is Hadoop, which is an open-source technology and initially had no security of any kind. Distributed and shared processing may point to the idea that less data would be processed by any of the system, yet it implies significantly more systems where security problems can manifest.

Computational security and other advanced resources in a distributed environment like MapReduce [50], a processing technique of Hadoop, tend to be deficient in security protection. MapReduce framework splits an input file into numerous lumps, and afterward, a mapper for each piece interprets the data, performs computations, and produces outputs as key-value pairs. Another actor, named reducer at that point, merges the values attached with every unique key and yields the outcomes. The chief fears here are: verification of the mappers and protection of data from a malevolent mapper. Mappers sending inaccurate outcomes are too tough to spot and inevitably bring about erroneous aggregate results. With very large data sets, malicious mappers are too hard to be detected as well, and they eventually damage essential data. Leakage of private records by mappers, be it intentional or accidental, is also an alarming matter. MapReduce calculations are frequently endangered by attack, such as man-in-the-middle, replay, and denial-of-service attack. Rogue nodes can be made part of the cluster and in order obtain replicated information or send modified MapReduce code. Making snapshots of authentic nodes and introducing their duplicate copies again is a simple attack in virtual environment and clouds, whereas it's difficult to detect it [32]. The two fundamental prevention approaches for it are to securely verify the mappers along with security of information within the sight of unauthorized and unlawful mapper.

4.11.2 NONRELATIONAL DATA STORES PROTECTION

Nonrelational databases, such as NOSQL databases adapted to store enormous volume of data, tackle several challenges associated with big data analytics without worrying a lot over security problems. NoSQL databases provide no clear security implementations, instead comprising security implanted in the middleware. Maintenance property transactional integrity is pretty negligent in these databases. Multifaceted integrity constrains can't be instilled in NoSQL datum as it hampers with its operational ability of giving improved scalability and performance. These databases have relatively weak authentication and password storing mechanisms. Its because they utilize basic or digest-related authentication (HTTP), and henceforth are vulnerable to the man-in-the middle attack. Also, representational state transfer is prone to injection attacks like JSON injection, array, view, REST, generalized query language, schema injection, and many more, including cross-site request forgery and cross-site scripting. NoSQL does not also support blocking with aid from third parties. Another limitation in terms of authorization approach is that it provides authorization only at higher layers. Consequently, it facilitates permission on each database level instead of providing it at data collection level. NoSQL databases are exposed to attack within it due to permissive security structures. Poor log analysis and logging methods leave these flaws unnoticed along with many other principal security mechanisms [32].

4.11.3 STORAGE SECURITY

Architecture of big data store data on multiple layers, conditional to business requirements for cost versus performance. For example, hot data of high precedence will typically be stored on smart flash medium. Therefore, confining the storage will be equivalent to creating tier-cognisant strategy.

A multitiered storage medium was used to keep transactional and data logs. The increase in data size, the issue of scalability, and availability led to the use of auto-tiering for storage of big data. But auto-tiering comes with a limitation: it does not record location of data, unlike past approaches of multitiering storage where the IT professional would exactly know where the data was and when was it placed there. This introduced new challenges for security of data storage. Just to state one, the service providers often look for hints that assist them to correlate user data sets and their activities and become acquainted with certain properties that can prove to be decisive for them. However, it's not possible for them to disrupt into data by conquering the encipherment. As the cipher text is stored by data owner in an auto-storage framework, he circulates the private key to every user and gives the right of data access to certain parts of the system to specific users. The unreliable service provider is not an authorized user of the key, but he might plot with other users by trading the key and data; subsequently, he can get access to data he is not legally allowed to access. If the environment is multitiered, service providers can promptly roll back an attack on set of users or deliver an obsolete form of data while the recent data is uploaded by then in the database. Data loss and data tempering came about by malicious users, regularly leading to conflicts among the users or between the providers [32].

4.11.4 MONITORING REAL-TIME SECURITY

Real-time security checking has been a progressing challenge in the case of big data analytics, primarily because of the amount of alerts security devices generated. These signals may or may not be correlated, but they can cause numerous false positives; because individuals lack the ability to effectively manage such an enormous amount of them at such pace, they are ignored or clicked away [51]. Monitoring of security necessitates that infrastructure or big data platforms be intrinsically secure. Several threats are posed to the infrastructure, which include web application threats, rogue administrative access to nodes or applications, and intrusion on the line. The security of each component of the infrastructure and security after their integration must be examined. Let's take the example of Hadoop, the cluster execute in a public cloud, hence the cloud security, which itself is an ecosystem of quite a few components comprising of storage, computing, and network mechanisms; it needs to be carefully studied. The security of Hadoop nodes, their interconnections, data storage, and the cluster in its entirety should be weighed. Also, monitoring application security containing pertinent associations that ought to pursue secure coding standards must be investigated as well, and the input sources from where the information originates must be accounted.

4.11.5 PRIVACY-PRESERVING DATA ANALYTICS AND MINING

Big data is vulnerable to privacy appropriation, decrease of civil freedom, obtrusive marketing, and an upsurge in corporate and state control. A worker in an organization responsible for storage of big data can possibly abuse power and disregard privacy guidelines. For instance: Big data workers can stalk individuals by observing conversations, if they work for a social networking-based company that aids chatting. An untrustworthy partner in business can invade personal data and transfer it into the cloud since policies permits the owner to handle cloud infrastructure [32].

4.11.6 GRANULAR AUDIT

The real challenge is when the monitoring system receives notification at the very instant an attack happens. There exists a possibility of frequent new attacks or unexploited true positives. To find a missed attack, required information is termed as audit information. Audit information from any gadget must be finished, or it must present us facts concerning what precisely occurred and what turned out wrong. It must deliver on-time access, with the goal that it obliges the need of compliance, forensic examination, rules, and regulations. It must not be meddled with and should be available only in permitted areas [32].

4.11.7 END-POINT SECURITY

Companies gather data from diverse sources, including software applications, hardware equipment, and end-point tools. Validation and the data collection from diverse sources is a serious challenge. Malicious users tamper with the target machine from where collection of data happened or temper the application responsible for collecting data, configuring the device to float malicious data as input to the core system of data collection. Counterfeit IDs are made by malignant users who intend to deliver malicious data as input to the data-collection system. Sybil attacks are a prominent ID cloning attack in a BYOD (bring your own device) setting where a vindictive user brings his own machine, forged as an authorized machine, and injects malicious input to the system. Sensory-data input sources can also be controlled, like artificially varying the temperature from its sensor and inserting malignant input into the collection process of temperature. Global positioning systems (GPS) can be altered similarly, where a user may modify the data during its transmission from source to the chief data-collection system. It can somehow be categorized under a man-in-the-middle attack [32].

4.11.8 DATA-CENTRIC SECURITY BASED ON CRYPTOGRAPHY

To control data visibility to individuals, systems, and organizations, two basic methodologies are commonly used. The first is to limit access to primary systems like hypervisor or operating systems. The next one is embodying the data in a defensive shield due to cryptography. The initial approach gives a bigger surface for attack. There exist numerous attacks, such as buffer overflow and privilege-based escalation attacks that circumvent access-control applications and gain data access. Shielding data from end to end using encryption support considerably reduces the attacking surface. Although it seems like an impossible task, it is susceptible to translate side attacks and fetch secret keys. Different threats related to cryptography-based, access-control strategy-exploiting encryption are: it ought not be recognizable by the enemy, and the equivalent plain-text data observes the cipher text, regardless of whether he needs to pick between right and wrong plain text. For a searching and refining of encrypted data, the cryptography protocol may not allow the adversary to learn anything regarding encrypted data past the relating predicate, regardless of whether fulfilled or not. It must also guarantee that the attacker must not have the option to construct information that originated from asserted sources; this probably would be bogus influencing data integrity [32].

4.12 SOLUTIONS TO SECURITY CHALLENGES

Several protections against security challenges have been proposed [52]. Few of theoretical solutions are discussed below.

4.12.1 Complete Data Supervision of Social Networks

In the era of big data, creation of online media has revolutionized interpersonal communication. Establishing strong supervision of data is critical. First, it is mandatory to fortify the management and supervision of data and secretly ensure the protection of data at network for mysterious social media. Second, conduct management and supervision of social data to safeguard private information security, and make sure this information is not exploited by the cyber criminal. Besides, to enhance users' knowledge of precautionary measures for safety and to mitigate personal information filling, vigilance drawbacks and self-prevention understanding are also required. At last, government bodies should announce better policies and regulation for big data applications at the earliest possible opportunity.

4.12.2 Improvement in Legal Mechanism

With the expansion of society, the public pays more attention to privacy, and governing bodies give more consideration to the safety of discrete rights of each citizen and present many counter measures for information protection. In the criminal law amendment, the principles for the defence of citizens' private information are stated explicitly; that is, come what may, the public officer sees about the resident's data, he/she must not utilize any resources to deliver information to others. If the information is leaked, it must accept some legal responsibility. Therefore, to protect the big data security, the administration needs to start a comprehensive personal information protection law to safeguard national's individual data.

4.12.3 Improvement to People Awareness of Data Quality

With the constant development in the era of big data, the amount of data has amplified naturally. People must adjust to variations in the times and progressively improve their data literacy and awareness. Data literacy is primarily designed for research scientists and public servants. Data awareness is focused at the overall public and need inhabitants to comprehend the significance of big data. Do not randomly print information regarding your own confidentiality on the internet, and do not publish others' information to avoid any exploitation by criminals.

4.12.4 Put Security First

While designing solution architecture, put security at the top-most priority. Being careful and aware of possible security risk at every stage allows IT professional to resolve the issues concurrently.

4.13 CONCLUSION

Big data came out from this incredible escalation in the number of IP-equipped end points. It's a fundamental term for all available data in any area that a corporation can collect with the aim of finding secret trends and patterns. These with the help of analytical tools can yield better results in the form of greater revenue and user satisfaction. Alternatively, the other side of the coin is that big data architecture, particularly its storage, depicts a new target of security issues for malware and criminal activities. Security and privacy are among the most talked about challenges of big data. Security breaches examples are there to showcase how data breaches can affect not only a firm's reputation but also its financial stability, along with sacrifice of users' confidential information. Proper policies are required to ensure security in big data, not only that users need to increase their awareness as well. Also, many data-mining and machine-learning algorithms, such as deep-learning models, can be exploited to reduce this problem.

REFERENCES

[1] Sagiroglu, S., & Sinanc, D. (2013). Big data: A review. In *2013 International Conference on Collaboration Technologies and Systems (CTS)* (pp. 42–47). IEEE.

[2] Almadhoor, L. (2021). Social media and cybercrimes. *Turkish Journal of Computer and Mathematics Education (TURCOMAT)*, 12(10), 2972–2981.

[3] Humayun, M. (2020). Role of emerging IoT big data and cloud computing for real time application. *International Journal of Advanced Computer Science and Applications*, 11(4), 1–13.

[4] Bekker, A. (2017). *Big data: Examples, sources, and technologies explained.* [online] Scnsoft.com. Available at: https://www.scnsoft.com/blog/what-is-big-data [Accessed 29Jul.2019].

[5] Humayun, M. (2021). Industry 4.0 and cyber security issues and challenges. *Turkish Journal of Computer and Mathematics Education (TURCOMAT)*, 12(10), 2957–2971.

[6] Zikopoulos, P., & Eaton, C. (2011). *Understanding big data: Analytics for enterprise class hadoop and streaming data.* McGraw-Hill Osborne Media.

[7] Alshammari, H. (2019). U.S. Patent No. 10,268,716. Washington, DC: U.S. Patent and Trademark Office.

[8] Laney, D. (2001). 3-D data management: Controlling data volume, velocity and variety. *META Group Research Note, 6*, 70.

[9] Dijcks, J. (2012). Big data for the enterprise. Oracle report.

[10] Schroeck, M., Shockley, R., Smart, J., Romero-Morales, D., & Tufano, P. (2012). Analytics: The real-world use of big data. IBM report, 1–20.

[11] Katal, W., & Goudar. (2013). Big data: Issues, challenges, tools and good practices, contemporary computing (IC3). In Sixth 2013 IEEE International Conference on Noida. IEEE.

[12] Krishnan, K. (2013). *Data warehousing in the age of big data.* Elsevier.

[13] Khurshid, K., Khan, A., Siddique, H., & Rashid, I. (2018). Big data-9Vs, challenges and solutions. *Technical Journal*, 23(03), 28–34.

[14] Lomotey, R. K., & Deters R. (2014). Towards knowledge discovery in big data. In *2014 IEEE 8th International Symposium on Service Oriented System Engineering.* IEEE.

[15] Boyd, D., & Crawford, K. (2012). Critical questions for big data: provocations for a cultural, technological, and scholarly phenomenon. *Information, Communication & Society,* 15(5), 662–679.

[16] Arockia Panimalar, S., Varnekha Shree, S., & Veneshia Kathrine, A. (2017). The 17 V's of big data. *International Research Journal of Engineering and Technology*, 4(9), 329–333.

[17] Shafer, T. (2019). The 42 V's of big data and data science. Available at: https://www.kdnuggets.com/2017/04/42-vs-big-data-data-science.html.

[18] George, G., Haas, M. R., & Pentland, A. (4 April 2014). Big data and management. *Academy of Management Journal*, 57(2). 10.5465/amj.2014.4002.

[19] Gill, N. S. (2017, Mar 03). Data ingestion, processing and architecture layers for big data and IOT. Available at: https://www.xenonstack.com.

[20] Sakr, S., & Gaber, M. (2014). *Large scale and big data: Processing and management*. Auerbach Publications.

[21] Kaul, L., & Goudar, R. H. (2016, November). Internet of things and big data-challenges. In 2016 Online International Conference on Green Engineering and Technologies (IC-GET) (pp. 1–5). IEEE.

[22] Lancaster, T. (2019, January 25). 9 Key big data security issues. Available at: https://www.alienvault.com/blogs/security-essentials/9-key-big-data-security-issues.

[23] Zaman, N., Seliaman, M. E., Hassan, M. F., & Márquez, F. P. G. (2015). *Handbook of research on trends and future directions in big data and web intelligence*. Pennsylvania: Information Science Reference.

[24] Oussous, A., Benjelloun, F. Z., Lahcen, A. A., & Belfkih, S. (2018). Big data technologies: A survey. *Journal of King Saud University-Computer and Information Sciences*, 30(4), 431–448.

[25] Nyunt, K., & Noor, Z. (2015). The effectiveness of big data in social networks. In *Handbook of research on trends and future directions in big data and web intelligence*, pp. 362–381. IGI Global.

[26] Tsai, C. W., Lai, C. F., Chao, H. C., & Vasilakos, A. V. (2016). Big data analytics. In *Big data technologies and applications,* pp. 13–52. Cham: Springer.

[27] Sivarajah, U., Kamal, M. M., Irani, Z., & Weerakkody, V. (2017). Critical analysis of big data challenges and analytical methods. *Journal of Business Research*, 70, 263–286.

[28] Ghosh, K., & Nath, A. (2016). Big data: Security issues and challenges. *International Journal of Research Studies in Computer Science and Engineering*, 1–9.

[29] Bertino, E., & Ferrari, E. (2018). Big data security and privacy. In *A comprehensive guide through the italian database research over the last 25 years,* pp. 425–439. Cham: Springer.

[30] Thayananthan, V., & Albeshri, A. (2015). Big data security issues based on quantum cryptography and privacy with authentication for mobile data center. *Procedia Computer Science*, 50, 149–156.

[31] Matturdi, B., Zhou, X., Li, S., & Lin, F. (2014). Big data security and privacy: A review. *China Communications*, 11(14), 135–145.

[32] Big Data Working Group. (2013). *Expanded top ten big data security and privacy challenges*. Cloud Security Alliance, 1–39.

[33] Abdrabo, M., Elmogy, M., Eltaweel, G., & Barakat, S. (2016). Enhancing big data value using knowledge discovery techniques. *IJ Information Technology and Computer Science*, 8, 1–12.

[34] Tene, O., & Polonetsky, J. (2011). Privacy in the age of big data: A time for big decisions. *Stanford Law Review Online*, 64, 63.

[35] Soria-Comas, J., & Domingo-Ferrer, J. (2016). Big data privacy: Challenges to privacy principles and models. *Data Science and Engineering*, 1(1), 21–28.

[36] Suganthi, M. (2018). Big data: Security issues, challenges and future scope. *International Journal for Research in Science Engineering & Technology*, 5(1), 10–20.

[37] Schmitt, C., & Shoffner, M. (2013). Security and privacy in the era of big data. *The SMW, a Technological Solution to the Challenge of Data Leakage*, 1(2).

[38] Kshetri, N. (2014). Big data's impact on privacy, security and consumer welfare. *Telecommunications Policy*, 38(11), 1134–1145.

[39] Cyber Edge Group. (2014). *2014 Cyberthreat Defense Report*. Cyber Edge Group.

[40] Alayda, S. (2021). Terrorism on dark web. *Turkish Journal of Computer and Mathematics Education (TURCOMAT)*, 12(10), 3000–3005.

[41] Mansfield-Devine, S. (2015). The Ashley Madison affair. *Network Security*, 2015(9), 8–16.

[42] Humayun, M., Niazi, M., Jhanjhi, N. Z., Alshayeb, M., & Mahmood, S. (2020). Cyber security threats and vulnerabilities: A systematic mapping study. *Arabian Journal for Science and Engineering*, 45(4), 3171–3189.

[43] Date, F. C., & Date, R. (2016). *Horizon Blue Cross Blue Shield of New Jersey*. https://www.horizonblue.com/sites/default/files/2017-05/horizon-bcbsnj-2016_annual_report.pdf.

[44] Ullah, A., Azeem, M., Ashraf, H., Alaboudi, A. A., Humayun, M., & Jhanjhi, N. Z. (2021). Secure healthcare data aggregation and transmission in IoT—A survey. *IEEE Access*, 9, 16849–16865.

[45] Thielman, S. (2016). Yahoo hack: 1bn accounts compromised by biggest data breach in history. *The Guardian*, 15, 2016.

[46] Trautman, L. J., & Ormerod, P. C. (2016). Corporate directors' and officers' cybersecurity standard of care: The Yahoo data breach. *American University International Law Review*, 66, 1231.

[47] Layton, R., & Watters, P. A. (2014). A methodology for estimating the tangible cost of data breaches. *Journal of Information Security and Applications*, 19(6), 321–330.

[48] Cadwalladr, C., & Graham-Harrison, E. (2018). Revealed: 50 million Facebook profiles harvested for Cambridge Analytica in major data breach. *The Guardian*, 17, 22.

[49] Terzi, D. S., Terzi, R., & Sagiroglu, S. (2015, December). A survey on security and privacy issues in big data. In 2015 10th International Conference for Internet Technology and Secured Transactions (ICITST) (pp. 202–207). IEEE.

[50] Almrezeq, N. (2021). "An enhanced approach to improve the security and performance for deduplication." *Turkish Journal of Computer and Mathematics Education (TURCOMAT)*, 12(6), 2866–2882.

[51] Singh, R., & Ali, K. A. (2016). Challenges and security issues in big data analysis. *International Journal of Innovative Research in Science, Engineering and Technology*, 5(1), 257–264.

[52] Zhang, D. (2018, October). Big data security and privacy protection. In *8th International Conference on Management and Computer Science (ICMCS 2018)*. Atlantis Press.

5 Prevention of DOS/ DDOS Attacks Through Expert Honey Mesh Security Infrastructure

Shehneela Khan, Tariq Ali, Umar Draz,
Sana Yasin, Muazzam A. Khan, and Amjad Ali

CONTENTS

DOI: 10.1201/9780367808228-5

5.1   INTRODUCTION

Network security is considered as the main component in information security because it is responsible for passing secure information between networks and saving
confidential information from unauthorized access. The need of the network security
system has been growing, along with the rapid increase of network users and their
transactions. Some network intrusion attacks can cause the unavailability of services
across networks. Such types of attacks are known as DOS attacks [1]. In client/server
environments, denial of service to the legitimate clients from the service providers
through the unavailability of services is the main problem. A modified version of
denial of service (DOS) attack is the distributed denial of service (DDOS) attack. In
DDOS attacks, intruders' attack from more than one location instead of a single one,
as shown in Figure 5.1. Intruders initially gain control over local systems. Such types
of infected systems are known as masters or handlers. Masters then interact with
further compromised systems that are closer to the target server. Combined effects of
attacks from several DDOS agents result in thousands of requests that overwhelm the
server. The server may slow down or crash in this situation [2]. DOS/DDOS attacks
are considered an important kind of cyberattack that may create various challenges
within the network environment. Following are the problem statements from an intensive review of literature and related work:

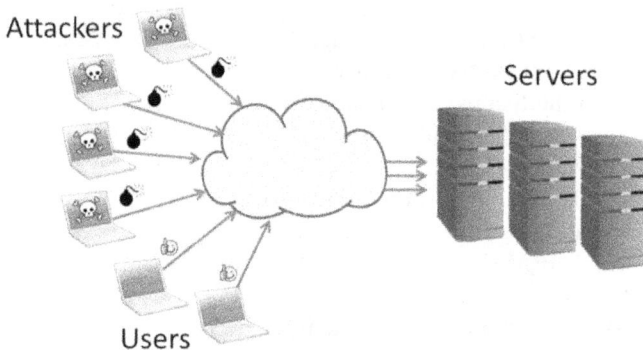

FIGURE 5.1   Existing possibilities between attackers and users.

1. The significant problem is when a group of people or automations attack a server or computer from many computers at once. This massive flow of data causes the resources of the server to end up not being enough, which causes the server to collapse and stop working [3]. The resources of the server computer are badly exhausted, i.e., increased CPU utilization, memory overload problems, and increased bandwidth utilization, etc.

2. Bandwidth is a useful resource within a computer network. Bandwidth utilization [4] is a more important factor, while transmitting information across the networks. When a DDOS attack is launched within a network, increased bandwidth utilization may slow down the traffic or data across the network.

3. It is necessary to know how to detect and prevent DOS/DDOS attacks [5] that cause security challenges across the network. Expert honey-mesh security infrastructure can be implemented to detect and prevent such kinds of attacks efficiently within the network.

The major objectives of this research are as follows:

- To implement expert honey-mesh security infrastructure within the network.
- To make fuzzy-logic rules and use expert honey-mesh systems to detect DDOS attacks within the network.
- To provide better bandwidth utilization and reduce the occurrence of DDOS attacks within the network.
- To conduct performance analysis of proposed security infrastructure through the software-simulation tool DDOSSim.

5.1.1 TYPES OF DDOS ATTACKS

The following are common types of DDOS attacks within the client/server architecture. These six types of DDOS attacks that can compromise the network traffic are observe in client/server environment [6]. Such types of attacks can compromise the network. These attacks are given below [7].

1. Protocol violation attacks
2. Fragmentation attacks
3. Network infrastructure attacks

5.1.1.1 Direct Flooding Attacks

These attacks are the simple version of DOS attacks. In this situation, an attacker directly attacks the victim's site by sending packets from his own system. The source address of the packets can be forged in this kind of attack. Many tools are available to allow this kind of attack for protocols such as TCP, UDP, and ICMP [8,9]. Some tools are commonly used, i.e., hping2, synsend, synk7, synhose, and stream2. In this situation, the attacker sent one packet, which is received by the targeted computer.

5.1.1.2 Remote-Controlled Network Attacks

In this kind of attack, attackers send an application within the targeted system and compromise the traffic within the networks. The computer listens to commands from the central computer. The network traffic can be compromised automatically or manually through a virus. Commonly used control channels are direct port communication, IRC channels, or ICMP ping packets. Zombie listens to TCP SYN packets passively across multiple destination ports in an organized way. A user-defined function [10,11] is called when ports are matched from any specific IP address. The attacker can use the packet header fields to get information about IP addresses and commands to run and attack the system. Compromised computers may launch the attacks directly to the target or through reflective media. It is really hard to trace the original control system in such type of attacks.

5.1.1.3   Reflective Flooding Attacks

In this situation, attackers forge the source IP address with the IP address of the victim and transmit them to an intermediate host. The intermediate host sends a reply to the destination of target computer [12,13]. It is difficult to recognize the actual attacker in reflective attacks because the packets are coming from inter-mediate servers.

5.1.1.4   Protocol Violation Attacks

The attacks by depletion of TCP state, also called attacks of protocol, are more sophisticated than the volumetric attacks and target some network devices, such as firewalls, web servers, and/or the load balancers [14,15]. The objective of the at-tacks by exhaustion of state is to exhaust the available resources of the target device and cause its failure.

5.1.1.5   Fragmentation Attacks

Packet fragmentation is divided into two regions, such as evasion of IDS detection and a DOS mechanism [16,17]. The system's resources are exhausted while re-assembling the packets during the fragmentation process that leads to DOS attack.

5.1.1.6   Network Infrastructure Attacks

Attacks always affect the overall operation of the internet and damage the whole network infrastructure. These attacks can cause global or regional network shutdowns or slowdowns. Recent attacks against root name servers over the in-ternet led to an FBI investigation. A warning signal will be sent to the root name servers' operators for the betterment of network infrastructure. Network outages are caused due to backbone services. They would include minimum extant RADIUS and DNS. Traffic within the network infrastructure can be divided into management plane, control plane, and data plane [18]. The data plane contains packets that may be sent to another place with the help of the router. The control plane consists of routing protocols that help the networks to work properly and efficiently. Network elements are managed through the tools and protocols the management plane addresses. Management and the control plane are forwarding to the router's processing engine. Router engines are not capable of handling large amounts of data that transmit via the data plane. They overwhelm CPU easily without any defense system [19].

5.2 WORKING OF DDOS ATTACKS

Network resources (such as web servers) have a finite limit of requests that they can attend at the same time. In addition to the server-capacity limit, the channel connecting the server to the internet has limited bandwidth or capacity. When the number of requests exceeds the capacity limits of any of the components of the infrastructure [20], the service level is likely to be affected in one of the following ways:

- The response to requests will be much slower than normal.
- Some (or all) of the users' requests may be ignored.

As a general rule, the primary intention of the attacker is to completely avoid the normal operation of the web resource, a total "denial" of the service. The attacker can also request a payment to stop the attack. In some cases, the objective of the DDOS attack [21] may be to discredit or damage a competitor's business.

5.2.1 USING A BOTNET "BOTNET" TO LAUNCH A DDOS ATTACK

To send an extremely large number of requests to the victim resource, the cyber-criminal often establishes a "zombie network" [22] of infected computers. As the offender controls the actions of each infected computer in the zombie network, the large scale of the attack can overwhelm the web resources of the victim.

5.2.2 PREVENTION MEASURES OF DDOS ATTACKS

Today, more than ever, business organizations are required to implement various security mechanisms to minimize an unauthorized access and to ensure the continuity of their services and activities. For most of the security plans, the key to the mitigation of DDOS attacks is to manage the threat before the execution of attacks [23].

The following recommendations help business organizations to reduce the risks related to DDOS attacks. One of the best ways to ensure that a company is in safe state is to make a plan of action against DDOS attacks. This plan of action should contain complete information of individuals and teams to contact in case of attack. Their roles and responsibilities should also be indicated in case the business suffers a DDOS attack.

An important aspect of this plan of action DDOS [24] should be the way in which you can handle with the current problem according to your customers, your staff, and other stakeholders of your business. Effective communication can help to control costs related to a DDOS attack and to build confidence in the people who matter the most for organizations. Your action plan DDOS should be checked and tested periodically to ensure that it covers every event and that it contains the information most relevant and up to date.

5.2.3 NEVER OVERESTIMATE THE DEFENSES OF THE NETWORK

The latest network strategies and security mechanisms have evolved considerably since the infrastructure organizations once used. The security mechanisms [25] should never be overestimated because attackers can attack if any vulnerability is found within the systems.

5.2.4 CREATE A REFERENCE MODEL TO BETTER IDENTIFY ACTIVE ATTACKS

It is irritating to note that some organizations are unaware of their network traffic levels in "normal" times. That's why they are also unaware if they suffer a DDOS attack. So there is a requirement to monitor the network traffic and defense reference models. They will thus be capable of determining if an increase of network traffic is the result of an attack or if it is simply the flow rate generally present at this specific period of time, the month or the year. It is also important to know where the majority of your network traffic is. If your company doesn't have clients and/or a network presence in a given country [26], but you found an important traffic from this country, the alarm should sound. The more quickly you recognize a DDOS attack, the greater are your chances of reducing the harm. However, low network performance and slow network traffic speed is not always due to DDOS attacks. That is why it is important to identify as quickly as possible the cause(s) first(s) of any degradation of service.

5.2.5 APPLY THE LATEST PATCHES OF SUPPLIERS

The updated servers and network devices with the suppliers' latest patches consist of critical function of any IT security plan, but that becomes particularly important in the fight against DDOS attacks [27]. The last thing you need when your organization is the victim of an attack is to discover that a supplier is reluctant to help you because you have not applied the latest patches. With the addition of new methods of DDOS attack always more accurate, known security flaws, for which patches exist, should be treated in priority.

5.2.6 SECURE THE IoT DEVICES

If your company is particularly visionary and seeks to seize the opportunities offered by the IoT devices [28–30], it must ensure that they are all saved against the remote access. Although the access to distance is sometimes important for troubleshooting by third parties, it also offers a gateway that makes it possible for hackers to infiltrate your network and to integrate your IoT [31] devices in a botnet world.

5.2.7 DEPLOY A SOLUTION FOR THE MITIGATION OF DDOS ATTACKS DEDICATED

For proper functioning, it is essential to deploy an efficient security system or proactive plan to mitigate DDOS attacks. Such kinds of security systems are capable of scanning suspicious activity or behavior and offer a greater serenity. For evaluation purpose [32], you must depend on your reference model that indicates the source stream, performance, time period, and other crucial aspects for the mitigation of attacks. Such types of solutions are associated with your stream of data. They send network traffic with centers of dedicated cleaning that filter the malicious traffic and allow only the authorized data to pass in real time. These kinds of systems must be a part of your overall plan of response to incidents so that the anomalies are quickly identified and resolved.

5.2.8   REDUCE THE ATTACK SURFACE

We should reduce the attacked surface area to minimize DDOS attacks, which can limit the options for attackers and gain protection in one place.

Protocols or applications that are not expected to be used for communication should not be exposed. It will help to reduce the expected points of attacks [33].

5.2.9   PLAN THE SCALING

There are two important things in reducing the large-scale volumetric DDOS attacks: the capability of servers and bandwidth (or transit) capacity to absorb and prevent attacks. In the case of a server's capacity, DDOS attacks can be volumetric attacks that exhaust the resources of victim server, i.e., RAM, disk, and increased CPU utilization, etc. It is important to reduce computing resources. To do this, it can generate greater computing resources or features, such as improved networks or broader network functions that support larger volumes. Moreover, the use of load balancers is also common to monitor or transfer loads from one resource to another resource to prevent overloading a resource [34].

When you develop your applications, ensure that your host providers give ample redundant internet connection in case of transit capacity.

It will help in managing large amounts of data or traffic. Since DDOS attacks stop the availability of continuous services or applications, it is necessary to put them close to large internet points of exchange.

It will help users to access the applications easily during high volume of internet traffic.

Know What Normal and Abnormal Traffic Is:

Whenever there is a high volume of internet traffic, the host will accept only the amount of data that it can easily manage without affecting the availability of services.

This is called the rate limit. Latest mitigation technique only accepts the authorized amount of data after analyzing the packets individually [35,36]. So, you need to gain accurate information about the data traffic the receiver usually receives. Moreover, many other techniques are already experimented with in our previous work [37–42].

5.2.10   IMPLEMENT FIREWALLS FOR SOPHISTICATED APPLICATION ATTACKS

It is good to use a web application firewall (WAF) against attacks of the type injection of SQL code. Moreover, in case of attack, you must be able to produce customized measures against malicious applications or coming from unexpected sources or invalid IP addresses, etc. To prevent attacks at the moment they execute, it can sometimes be useful to take special support to analyze traffic patterns and produce the personalized protective measures. As users, we are required to review the configuration of our firewalls or routers to detect false IPs coming from expected attackers. Normally, our internet service provider (ISP) ensures that our router is up to date with this configuration. On the other hand, organizations that offer these services must save both their network and their whole infrastructure to mitigate such attacks from affecting the performance of their work and, consequently, to their clients [43]. If a company is affected by a denial-of-service (DOS) attack, it will lose the trust of its clients and lose customers for its services.

5.3 METHODOLOGY

DDOS attacks cause the unavailability of services to the legitimate users. They start from overall analysis of the work. In DDOS attacks, intruders or attackers execute attacks from multiple locations rather than a single one. Some local systems re controlled by the attacker. The attacker initially compromises and gains control over these local systems. Such a compromised system is known as a master or handler. Masters are then used to gain control over further compromised systems that are closed to the victim server. These systems are called slaves or DDOS agents. They further launch multiple attacks on victim servers, as shown in Figure 5.2.

5.3.1 INTRODUCTION TO EXPERT HONEY MESH SYSTEM

Hackers are always in search of vulnerabilities within the systems for the execution of attacks. Systems that are not sufficiently protected can be easy targets for

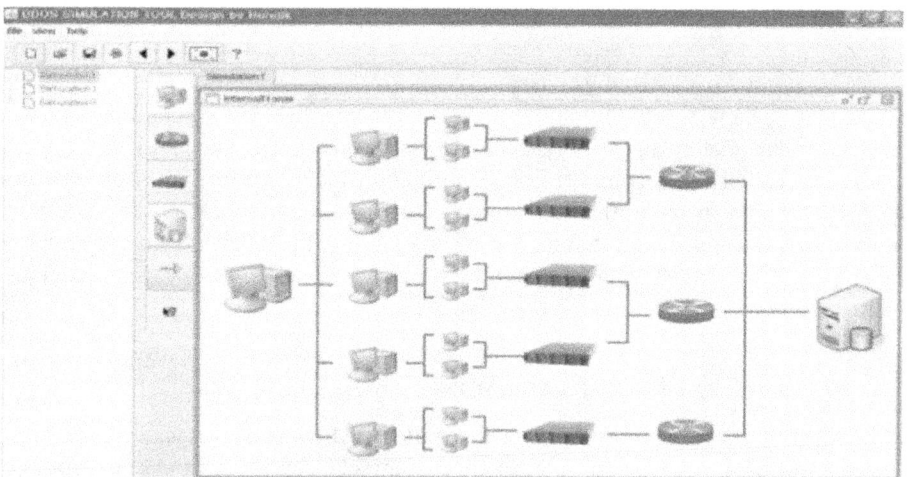

FIGURE 5.2   DDOS attacks execution.

hackers. In computer terms, a honey pot is a security mechanism with which administrators trap the hackers and cyberattacks are executed in vain [10]. A honey pot offers network services or application programs that attract the hackers and save the production system against possible damage. In practice, users gather server-side or client-side technologies to create honey pots [12].

5.3.2 CREATION OF HONEY POTS

Honey pots can be created on both sides as given below:

A. Creation of honey pots on the server side
The basic idea of a honey pot on the server side is to lure attackers to separate parts of a computer system and keep them away from production servers within the networks.
B. Creation of honey pots on the server side
A honey pot on the client side imitates application software that makes use of services on the server.

5.3.3 IMPLEMENTATION OF HONEY POT

Honey pot can be implemented in three zones.

A. In Front of the Firewall
By placing it in front of the firewall, it ensures that the security of our internal network is not compromised at any time as our firewall will prevent the attack from going to our internal network.
B. Behind the Firewall
This option allows us to control internal and external attacks of any kind; the main problem that this method presents is that it requires a specific configuration to leave access to the *honey pot,* but not to our network, which causes possible security failures in the traffic filtration.
C. In a Demilitarized Zone
By positioning it here, it is possible to separate the *honey pot* from the internal network and the union with the servers, and this possibility allows us to detect both internal and external attacks with a small modification of the Firewall.

5.4 PROPOSED SYSTEM

In this research, expert honey-mesh systems are used for the prevention of DDOS attacks within the client-server environment. Our expert honey-mesh system has features that ensure better security for the organization. It's very useful for the security of the information, since, through the detailed study with the captured information of the attackers, new means and techniques can be created against the attacks of hackers. Although the use of honey pots or honey nets is of great value, one should never entrust the security of the organization to such tools alone; they are not direct means of protection, but rather, means of study. They must work together with conventional preventive means to prevent an attacker from attacking other systems or networks. So, we have added fuzzy-logic rules in our proposed expert honey-mesh system to prevent DDOS attacks efficiently. Our proposed expert honey-mesh security infrastructure is used to detect and prevent DOS/DDOS attacks by using fuzzy-logic rules. The proposed methodology is very flexible and protects the production server from the malicious requests of attackers. In this research, honey-mesh systems are implemented before the firewall and are used to trap the attacker. Detection and prevention of DDOS attacks through an expert honey-mesh system will be based on fuzzy-logic rules. Performance of security systems will be evaluated through software simulation tools DDOSSIM. MATLAB® 10.1 is used to generate fuzzy-logic rules against DDOS attacks. These rules are used to detect attacks efficiently.

5.4.1 WORKING OF PROPOSED SYSTEM

Figure 5.3 illustrates the working of our proposed honey-mesh security infrastructure:

5.4.1.1 Detection and Prevention of DDOS Attack Via Our Proposed System

In our proposed system, an attacker sends a malicious stream of requests toward honey VM or honeyed. Honey VM has fuzzy rules about an expected traffic for analyzing

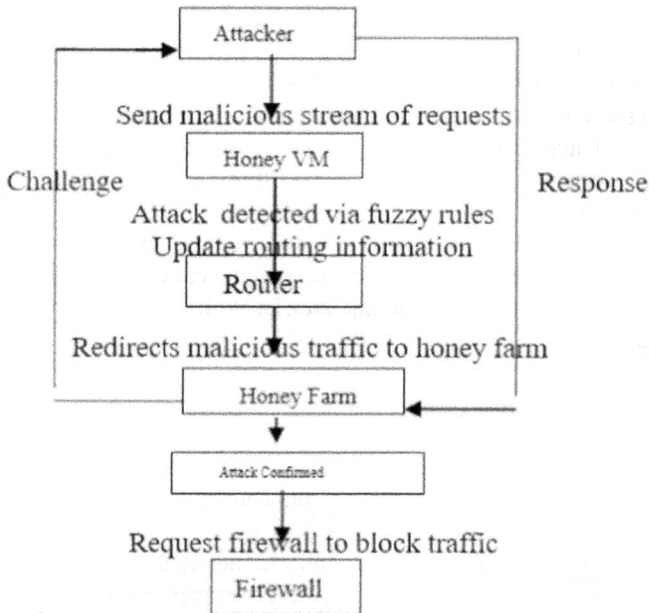

FIGURE 5.3 Proposed expert honey-mesh system.

few requests. Fuzzy rules are used to detect the malicious requests from incoming traffic; if any change is noticed, further investigation is required to determine whether the request consists of DDOS attacks. Once the honey VM detects a malicious request based on fuzzy rules or fuzzy reasoning, it requires further verification of whether it is an actual DDOS attack. It will send this updated information to the router. The router will route this updated information toward the honey farm. In this way, malicious traffic will not move toward the production server.

5.5 EXPERIMENTAL RESULT

Attackers launch DDOS attacks from multiple locations by using different compromised systems. They can be effective and exploit computers and other network resources, such as IOT devices. DDOS attacks can overwhelm the RAM and hard drives, and network bandwidth is badly affected in case of DDOS attacks. It also causes increased network congestion. Network congestion creates problems for router-to-router information across the network that leads to DDOS attacks. Due to increased traffic across the network, servers will not be able to perform their normal activities. So, they cannot fulfil client requirements efficiently. Such kinds of attack overwhelm the web server of our website.

5.5.1 SIMULATION

We have used simulation software DDOSIM to simulate the real scenario. Following are the figures that show the two cache servers, i.e., web server and file server. A graph is drawn between the number of packets and time in seconds. The graph shows x and y components to represent time in seconds and packets in GBS. The X-axis shows time (seconds), whereas y-axis shows packets in GBS. Figure 5.4 shows the

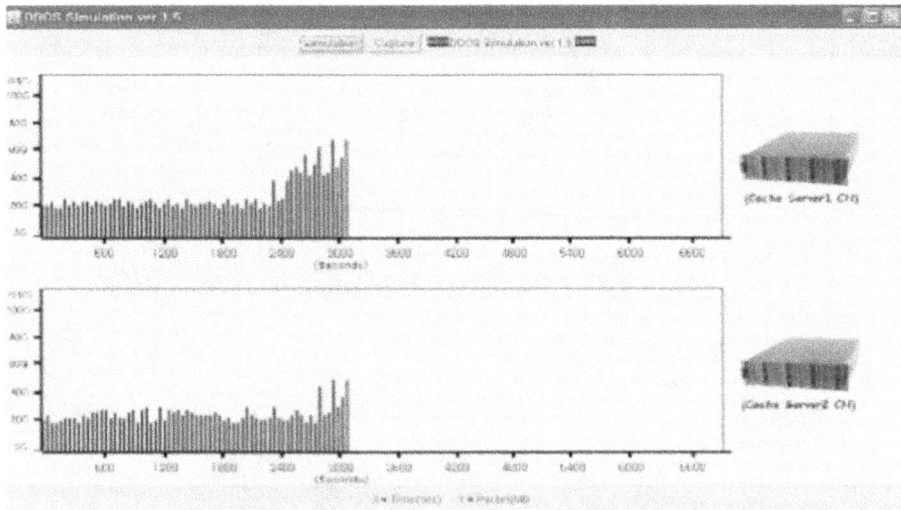

FIGURE 5.4 Bandwidth without DDOS attack.

data-transfer rate or data volume is normal when there is no DDDOS attack. Cache server 1 receives 20 GBS packets between 600 seconds and 2300 seconds. Packets are increased from 40 GBS to 80 GBS between 2400 seconds and 3000 seconds when the attack executes. Cache server 2 receives 20 GBS packets between 600 seconds to 2700 seconds. Packets are increased from 20 GBS to 40 GBS from 2900 seconds. In this situation, before DDOS attack, the bandwidth is normal, but after attack, bandwidth is increased in both servers, which exhausts the resources of both servers. It may slow down the actual traffic.

The resources of victim server result in the unavailability of services to legitimate users. They can also increase the network bandwidth that may slow down the actual traffic. So there is a need of an efficient security system to detect and prevent DDOS attacks within the network.

5.5.2 DDOS ATTACK LAUNCHED ON WEBSITE

An attack is launched on website www.neeosearch.com from multiple locations through compromised local systems known as masters or handlers that coordinate with agents. Agent systems are actual attack executers. Now attack is executed within an organization that results in increased network bandwidth. CPU utilization is also increased, and system resources like CPU, hard drives, cache server, and RAM are also exhausted. Network bandwidth and system resources are badly exhausted. Network congestion has also occurred, which wasted the actual traffic results caused by DDOS attacks.

5.5.3 INCREASE IN THE VOLUME OF ATTACKS AND DISTRIBUTION OVER TIME

When an attack is launched over www.neeosearch.com, network bandwidth is increased due to flooding of requests. In DDOS attacks, many devices are coordinated for the purpose of transmitting large volumes of illegitimate traffic toward the victim. In case of DDOS attack execution, network bandwidth is increased that cause the server's resources to be exhausted due to increased CPU utilizationresources like shown in Figure 5.5.

System diagram shows the name of the system in such type of inference system (Figure 5.6).

Input Variables
  Input 1= Victim's MTU (Maximum transmission unit) Input 2= Packet length of ICMP message
  Input 3= 'Hop count'
  Output Variables
  Output1= Volumetric attack Output2=ICMP flood attack Output3= DDOS attack

---

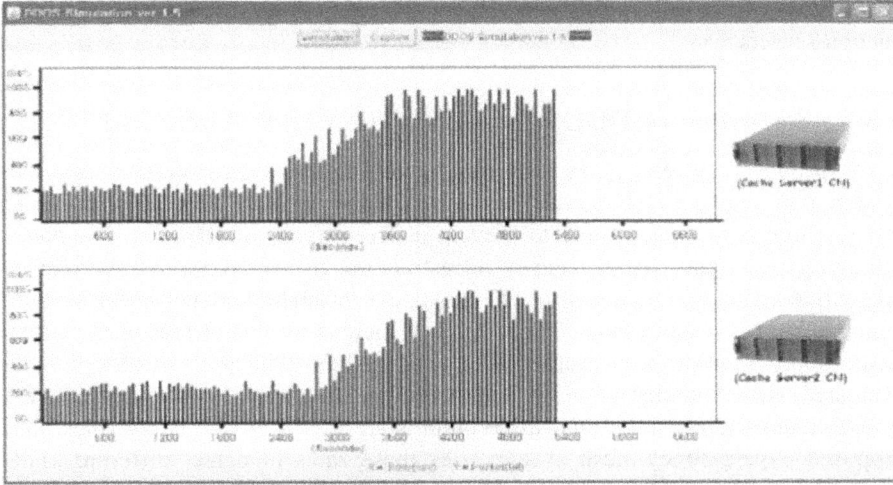

FIGURE 5.5   Bandwidth during DDOS attack.

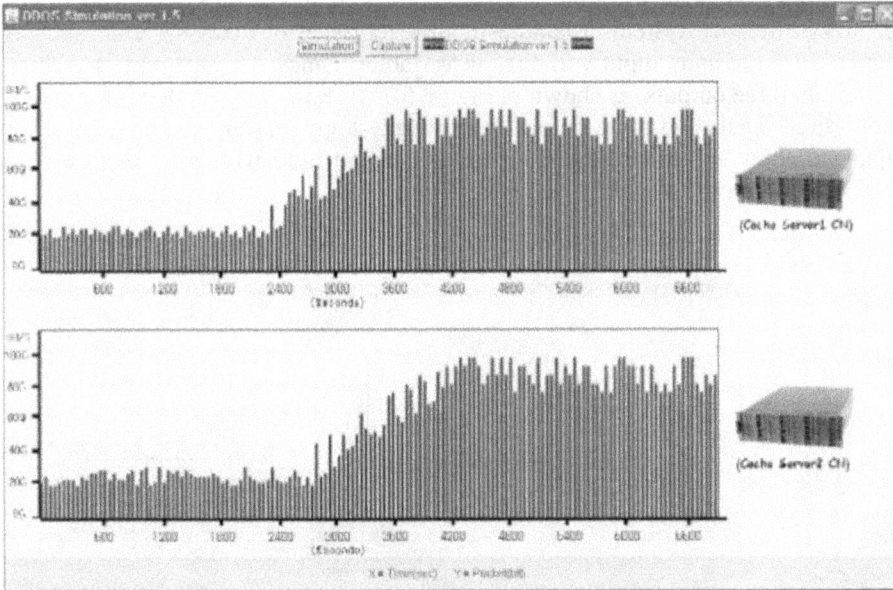

FIGURE 5.6   Bandwidth consumes server resources, i.e., cache server.

In our proposed solution, where a honey-mesh system is implemented before the firewall, it can efficiently detect DDOS attack through fuzzy-logic rules and prevent them efficiently through a firewall. The honey-mesh system observes the incoming traffic and notices the malicious activities continuously; once an attack is detected via applied **fuzzy-logic rules**, all traffic from attack source will be forwarded to the honey VM or honeyed. In this way, malicious data will not route toward the production servers.

5.5.4 PREVENTION OF DDOS ATTACK VIA FUZZY LOGIC MECHANISM

In our proposed system, fuzzy reasoning is only used by the expert honey-mesh system to prevent the DDOS attacks detected by the systems' VM. The purpose of this research is to explain how to prevent different types of DOS/DDOS attacks through expert honey-mesh security infrastructure. In our proposed solution, the preventive action will be taken for DDOS attacks through the honey-mesh system, which is based on fuzzy rules. Fuzzy control logic is used with reasoning, which is the approximation value rather than exact values. The performance will be evaluated via taxonomy of input and output parameters to detect and prevent DDOS attacks. Three input and three outputs are used to make fuzzy rules. Our proposed expert honey-mesh system uses these rules to detect different kinds of DDOS attacks. Every input has three membership functions. These are the inputs with their membership functions: MATLAB 10.1 software is used to create such rules. The fuzzy-logic toolbox provides a graphical user interface by typing fuzzy in the command window. The FIS editor will appear on screen. FIS stands for fuzzy inference system. It handles high-level issues of fuzzy systems, such as number of input and output variables in variables names. This example has three inputs and three outputs, as shown in Figure 5.7. It shows the detailed information about three inputs and three outputs using fuzzy-logic reasoning. There are three input variables and three output variables that are used with their membership functions within the system.

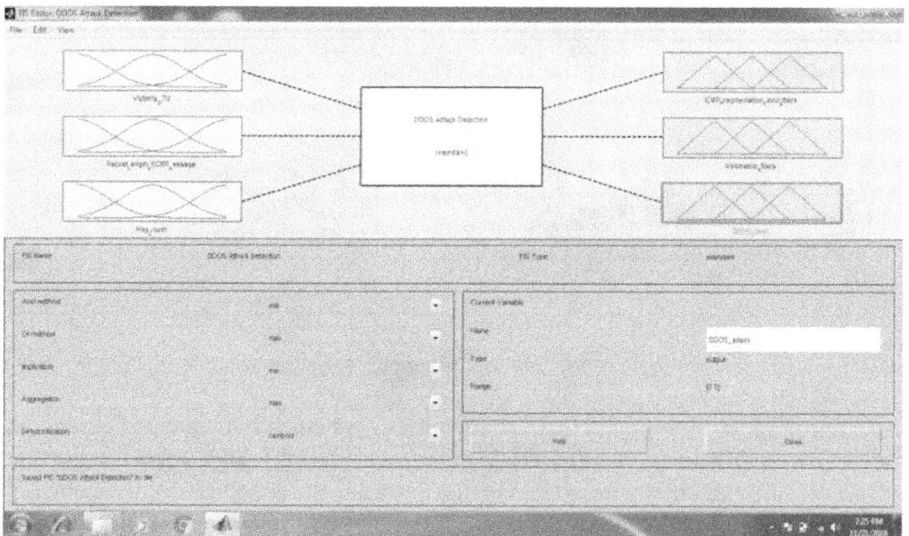

FIGURE 5.7   Inputs and outputs for detection of DDOS attacks.

Input 1= Victim's MTU (Maximum transmission unit)

Packet length of connection request message to the receiving network (receiver's MTU, i.e., usually 1500 bytes).

Membership functions of input 1

Input 1< 1500 bytes less fragmented packets Input 1 = 1500 bytes normal fragmented packets Input1> 1500 bytes high fragmented packets

Logic = attackers send highly-spoofed, large fragmented ICMP packets (1500+ byte) at a very high packet rate and such types of packets are difficult to reassemble. If packet size is large, it may enlarge the bandwidth of an ICMP attack overall. It exhausts CPU resources in an attempt to reassemble the wasted and useless packets. If Input 1, i.e., MTU is less than 1500 bytes, then packets will be less fragmented. If input 1, i.e., MTU is equal to 1500 bytes, then packets will be normal fragmented. If Input 1, i.e., MTU is greater to 1500 bytes, then packets will be highly fragmented.

Input 2= packet length of ICMP message

i.e. ICMP message of victim's spoofed source IP address (The minimum size of any ICMP packet, or message, is much lower than the minimum ethernet frame size on an ethernet network, which is 64 bytes.)

Membership functions Input2>64 bytes large ICMP packet size Input2=64 bytes normal ICMP packet size input 2<64 bytes less ICMP packet size

Logic: Attackers send large numbers of ICMP packets with the intended victim's spoofed source IP address and they are routed to a computer network using an IP broadcast address. This causes all hosts on the network to reply to the ICMP request, causing large volume of data to the victim computer. (Dao, Kim et al. 2016) If input 2 is less than 64 bytes, it will result in less ICMP packet size. If input 2 is equal to 64 bytes, it will result in normal ICMP packet size. If input 2 is greater than 64 bytes, it will result in large ICMP packet size.

Input 3 Hop Count Range= Default TTL value= 255 If Input 3 < 255 less hop limit

Input 3 = 255 normal hop limit Input 3 > 255 greater hop limit

Logic: If hops are more than 255, it may lead to increased CPU utilization and system instability. With the help of void IP addresses, the attacker starts sending mass RREQ messages toward the destination. It will send excessive RREQ without checking the RREQ HOPLIMIT, i.e., without waiting for the ROUTE REPLY. Besides, in the beginning, the TTL of RREQ is set up to a maximum without using an expanding ring-search process. In this situation, the whole network will be full of RREQ packets from the attacker that leads to full attack (Jung 2015). If input 3 is less than 255 hops, it will result in a smaller hop limit. If input 3 is equal to 255 hops, it will result in normal hop limit. If input 3 is greater than 255 hops, it will result in greater hop limit.

Output Variables Output1 Fragmented Flood Attack

(Membership Functions of Fragmented Flood Attack)

Less fragmented flood attack Normal fragmented flood attack High fragmented flood attack

If input 1, i.e., MTU is less than 1500 bytes, less fragmented packets will be generated that lead to less ICMP flood attack. Output will be less fragmented packets.

If input 1, i.e., MTU is equal to 1500 bytes, normal fragmented packets will be generated that lead to no ICMP flood attack. Output will be 'NO ICMP flood attack'.

If input 1, i.e., MTU is greater to 1500 bytes, more fragmented packets will be generated that lead to high ICMP flood attack. Output will be 'High ICMP flood attack'.

Output2 Volumetric Attack (Membership Functions of Volumetric Attack)

Less volumetric attack No volumetric attack High volumetric attack

If input 2, i.e., packet length of ICMP message is less than 64 bytes, smaller of ICMP packets will be generated that lead to less volumetric attack. Output will be 'less volumetric attack'. If input 2, i.e., packet length of ICMP message is equal to 64 bytes, normal size of ICMP packets will be generated that lead to no volumetric attack. Output will be 'no volumetric attack'. If input 2, i.e., packet length of ICMP message is greater than 64 bytes, greater size of ICMP packets will be generated that lead to high volumetric attack. Output will be 'High volumetric attack'.

Output 3 DDOS Attack (Membership Functions of DDOS Attack)

Less DDOS attack No DDOS attack High DDOS attack

If input 3, i.e., hop count is less than 255 hops, less hop limit will be generated that lead to less DDOS attack. Output will be 'Less DDOS attack'. If input 3, i.e., hop count is equal to 255 hops, normal hop limit will be generated that leads to no DDOS attack. Output will be 'no DDOS attack'. If input 3, i.e., hop count is greater than 255 hops, greater hop limit will be generated that leads to high DDOS attack. Output will be 'high DDOS attack'.

5.5.4.1   Fuzzy-Logic Rules

1. If (input1 is Less_Fragmented_Packets) or (Packet_Length_of_ICMP_message is Less_ICMP_Packet_Size) or (Hop_Count is Less_hop_limit) then (ICMP_Fragmentation_Flood_attack is Less_ICMP_flood_attack)(Volumetric_Attack is Less_Volumetric_attack)(DDOS_attack is Less_DDOS_attack)

2. If (input1 is Normal_Fragmented_Packets) or (Packet_Length_of_ICMP_message is Normal_ICMP_Packet_Size) or (Hop_Count is Normal_hop_limit) then (ICMP_Fragmentation_Flood_attack is No_ICMP_flood_attack)(Volumetric_Attack is No_Volumetric_attack)(DDOS_attack is No_DDOS_attack) (1)

3. If (input1 is High_Fragmented_Packets) or (Packet_Length_of_ICMP_ message is High_ICMP_Packet_Size) or (Hop_Count is Greater_hop_limit) then (ICMP_Fragmentation_Flood_attack is High_ICMP_flood_attack) (Volumetric_Attack is High_volumetric_attack)(DDOS_attack is HIgh_ DDOS_attack) (1)

5.5.5 Implementation of Fuzzy-Logic Rules for Prevention of DDOS Attacks

Honey-mesh systems use the above-mentioned fuzzy-logic rules to trap or detect DDOS attacks when traffic is passed on the website www.neeosearch.com. Now, DDOS attack is launched again on website www.neeosearch.com from various locations through compromised systems. In this situation, CPU utilization is increased, which consumes more network bandwidth and exhausts the server's resources. When the attacker executes a DDOS attack, he sends a malicious request to the server of the website. Honey-mesh systems are continuously observing the traffic that is passing toward the actual server of the website. Honey VM is installed within our proposed expert honey-mesh system that will help to prevent DDOS attack. Our defined fuzzy-logic rules are implemented within honey VM. Honey VM analyzes the traffic according to fuzzy rules and detects the DDOS attack. When an attack is detected, it updates this routing information to the router. Router redirects this information toward the honey farm VM where it sends challenging query in the form of puzzle or mathematic question to the attacker. Attacker gives response to this challenging query from honey-farm VM. On his response, honey VM can easily detect whether it is machine or human intervention. When DDOS attack is confirmed by the honey VM, it sends request to the firewall to discard or block that malicious traffic. In this way, our proposed expert honey-mesh system detects and prevents DDOS attack and saves the actual physical server of the website www.neeosearch.com. We have used simulation software DDOSSIM to simulate the whole process efficiently. Following are the results that have been taken from the simulation environment. For simulation, DDOSSIM simulation software is used. Figures 5.8–5.10 shows the data-transfer rate or data volume is normal after passing the traffic to an expert honey-mesh system. Cache server 1 receives 20 GBS packets between 600 seconds and 2300 seconds before the execution of DDOS attack. Packets are increased from 20 GBS to 70 GBS between 2400 seconds and 3000 seconds after the execution of the DDOS attack. When this traffic is passed within our proposed expert honey-mesh system, bandwidth will be decreased from 40 GBS to 20 GBS back within the given amount of time. Now the server's resources are free for actual traffic. Cache server 2 receives 20 GBS packets between 600 seconds and 2700 seconds before the execution of the DDOS attack. Bandwidth is normal. Packets are increased from 20 GBS to 50 GBS within the given amount of time, i.e., from 2800 seconds to 4000 seconds. When this traffic is passed within our proposed expert honey-mesh system, bandwidth will be decreased from 40 GBS to 20 GBS back within the given amount of time. Now server's resources are free for actual traffic.

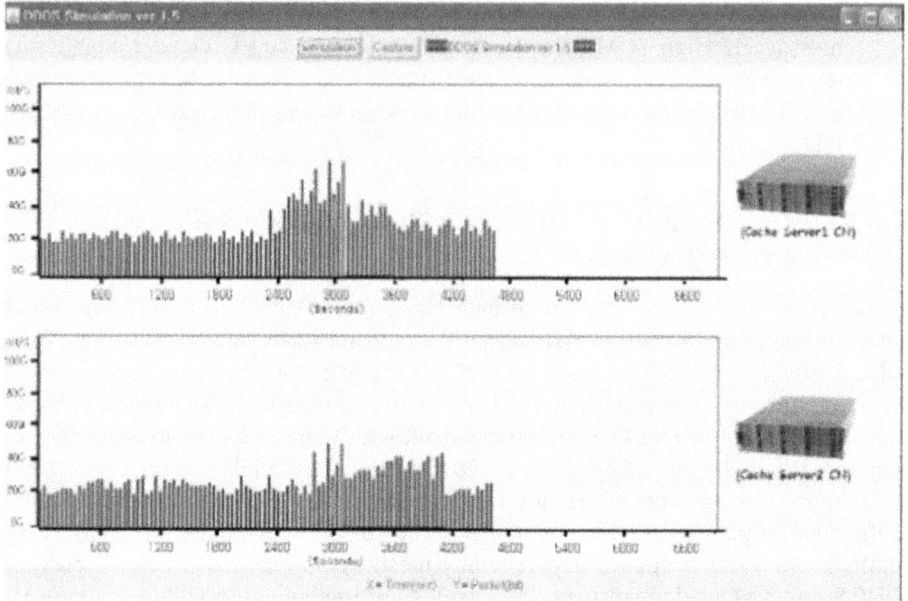

FIGURE 5.8 Bandwidth after passing traffic to expert honey mesh system.

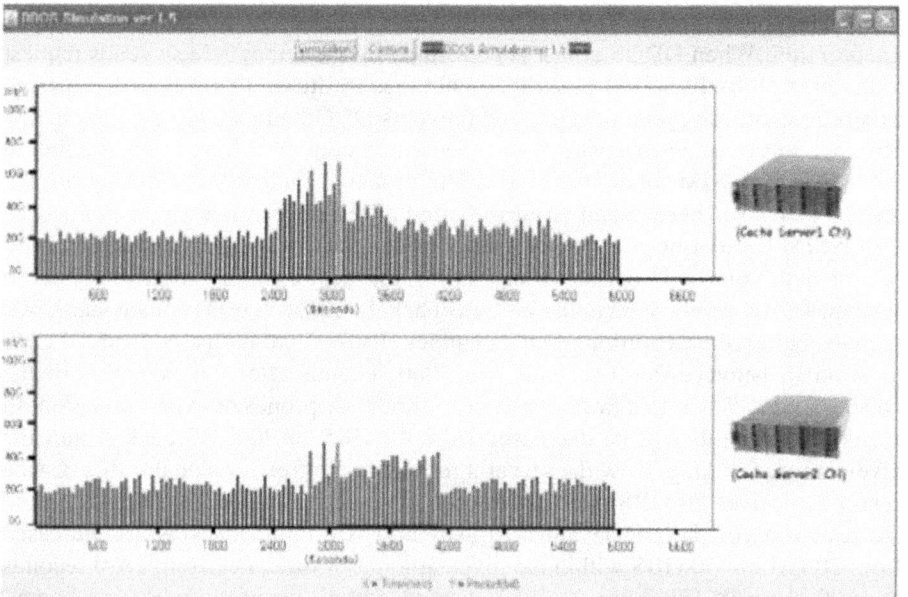

FIGURE 5.9 Bandwidth after traffic passing to an expert honey mesh system.

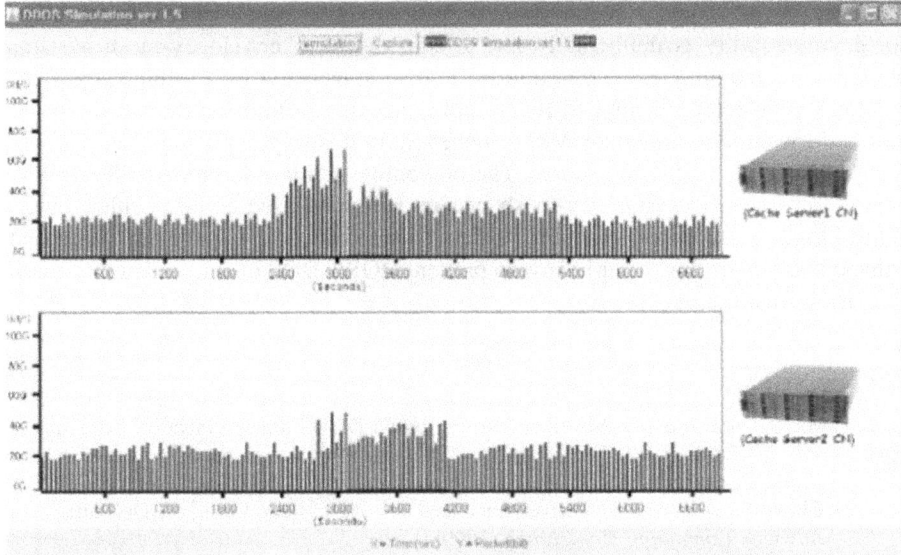

FIGURE 5.10   Bandwidth after traffic passing to an expert honey mesh system.

5.5.6 ADVANTAGES AND FUTURE ENHANCEMENTS

In this research, the proposed solution of preventing DDOS attacks via an expert honey-mesh security system have many advantages over existing solutions.

Following are advantages of expert honey-mesh security systems:

- The proposed solution is economical and has low-maintenance overheads due to the implementation of VM's rather than actual physical servers.
- In the proposed security solution, an additional VM is used as a backup. It ensures the process of instruction detection and prevention continuously in case an existing VM is attacked or compromised by the attacker.
- The process of restoring a compromised VM is cheap and takes less time.
- The proposed solution is in static mode due to predefined fuzzy rules, but we can extend it in dynamic mode in future in case of changed behavior of attacker.

5.6 CONCLUSION

DDOS attacks are used to continue the unavailability of services to legitimate users within networks. Expert honey-mesh systems are used to ensure the continuous availability of services across networks. Honey pot may be defined as a trap that mimics, notices, and records overall activities of the attacker. Low-interaction honey pot is used on the server side, recognizes different kinds of attacks, protects users' confidential data, and records all malicious activities via fuzzy-logic rules.

The security mechanism of honey-mesh systems is the same as real servers, but some vulnerability is deliberately left to trap the attacker. Honey-mesh systems observe the incoming traffic and notice the malicious activities continuously; once an attack is detected via applied fuzzy-logic rules, all traffic from the attack source will be forwarded to the honey VM or honeyed. In this way, malicious data will not route toward the production server. The preventive action will be taken for DDOS attacks through an expert honey-mesh system, which is based on fuzzy rules. Fuzzy control logic is used with reasoning, which is approximation value rather than exact values. The aim of this paper is how to prevent DOS/DDOS attacks via honey-mesh security infrastructure.

REFERENCES

[1] He, Z., et al. (2017). Machine learning based DDOS attack detection from source side in cloud. In *2017 IEEE 4th International Conference on Cyber Security and Cloud Computing (CSCloud)*.

[2] Chow, S., et al. (2009). Performance analysis of a system during a DDoS attack. In *Advance Computing Conference, 2009*.

[3] Osanaiye, O., et al. (2016). Distributed denial of service (DDoS) resilience in cloud: Review and conceptual cloud DDoS mitigation framework. *Journal of Network and Computer Applications*,67, 147–165.

[4] Beitollahi, H., and Deconinck, G. J. C. C. (2012). Analyzing well-known countermeasures against distributed denial of service attacks. *Computer Communications*, 35(11), 1312–1332.

[5] Deshpande, H. A. (2015). HoneyMesh: Preventing distributed denial of service attacks using virtualized honeypots. arXiv.org.

[6] Humayun, M. (2021). Industry 4.0 and cyber security issues and challenges. *Turkish Journal of Computer and Mathematics Education (TURCOMAT)*, 12(10), 2957–2971.

[7] Mousavi, S. M., and St-Hilaire, M. (2015). Early detection of DDoS attacks against SDN controllers. In *2015 International Conference on Computing, Networking and Communications (ICNC)*.

[8] Jaber, A. N., et al. (2017). Methods for preventing distributed denial of service attacks in cloud computing. *Advanced Science Letters*, 23(6), 5282–5285.

[9] de Almeida, M. P., et al. (2018). New DoS defense method based on strong designated verifier signatures. *Sensors (Basel, Switzerland)*, 18(9), 2813.

[10] Hussain, K., Hussain, S. J., Jhanjhi, N. Z., and Humayun, M. (2019). SYN flood attack detection based on Bayes estimator (SFADBE) for MANET. In 2019 International Conference on Computer and Information Sciences (ICCIS), pp. 1–4. IEEE.

[11] Prestele, A. Middleboxes Against DDoS Attacks.

[12] Sokol, P., et al. (2017). Honeypots and honeynets: Issues of privacy. *EURASIP Journal on Information Security*, 2017(1), 4.

[13] Sachdeva, M., et al. (2010). DDoS incidents and their impact: A review. *The International Arab Journal of Information Technology*, 7(1), 14–20.

[14] Gopi, R., Sathiyamoorthi, V., Selvakumar, S., Manikandan, R., Chatterjee, P., Jhanjhi, N. Z., and Luhach, A. K. (2021). Enhanced method of ANN based model for detection of DDoS attacks on multimedia internet of things. *Multimedia Tools and Applications*, 1–19.

[15] Mirkovic, J., and Reiher, P. (2004). A taxonomy of DDoS attack and DDoS defense mechanisms. *ACM SIGCOMM Computer Communication Review*, 34(2), 39–53.

[16] Jain, A., et al. (2015). Advance trends in network security with Honeypot and its comparative study with other techniques. *International Journal of Engineering Trends and Technology*, 29, 304–312.

[17] Wang, C., et al. (2017). Research on DDoS attacks detection based on RDF-SVM. In *2017 10th International Conference on Intelligent Computation Technology and Automation (ICICTA)*.

[18] Gasti, P., et al. (2013). DoS and DDoS in named data networking. In *2013 22nd International Conference on Computer Communication and Networks (ICCCN)*.

[19] Almadhoor, L. (2021). Social media and cybercrimes. *Turkish Journal of Computer and Mathematics Education (TURCOMAT)*, 12(10), 2972–2981.

[20] Kumarasamy, S., and Asokan, R. J. (2012). Distributed denial of service (DDoS) attacks detection mechanism. *International Journal of Computer Science Engineering and Information Technology*, 1(5).

[21] Nezhad, S. M. T., et al. (2016). "A novel DoS and DDoS attacks detection algorithm using ARIMA time series model and chaotic system in computer networks. *IEEE Communications Letters*, 20(4), 700–703.

[22] Alomari, E., et al. (2012). Botnet-based distributed denial of service (DDoS) attacks on web servers: Classification and art. *Shadow Honeypots,* 2(9), 1–16.

[23] Prasad, K. M., et al. (2014). DoS and DDoS attacks: Defense, detection and traceback mechanisms-a survey. *Global Journals*, 14(7).

[24] Mahajan, D., and Sachdeva, M. (2013). DDoS attack prevention and mitigation techniques: A review. *International Journal of Computer Applications*, 67(19).

[25] Sharma, N., et al. (2011). Attack prevention methods for DDOS attacks in MANETs. 1(1), 18–21.

[26] Singh, B., et al. (2015). An adaptive approach to mitigate DDoS attacks in cloud. *International Journal of Advanced Computer Science and Applications*, 6(10).

[27] Wang, C., et al. (2018). Skyshield: A sketch-based defense system against application layer DDoS attacks. *IEEE Transactions on Information Forensics and Security*, 13(3), 559–573.

[28] Batool, S., Saqib, N. A., and Khan, M. A. (2017, May). Internet of Things data analytics for user authentication and activity recognition. In 2017 Second International Conference on Fog and Mobile Edge Computing (FMEC), pp. 183–187. IEEE.

[29] Jhanjhi, N. Z., Brohi, S. N., Malik, N. A., and Humayun, M. (2020). Proposing a hybrid RPL protocol for rank and wormhole attack mitigation using machine learning. In 2020 2nd International Conference on Computer and Information Sciences (ICCIS), pp. 1–6. IEEE.

[30] Ali, T., Khan, M. A., Hayat, A., Alam, M., and Ali, M. (2013). Secure actor directed localization in wireless sensor and actor networks. *International Journal of Distributed Sensor Networks*, 9(10), 126327.

[31] Batool, S., Saqib, N. A., Khattack, M. K., and Hassan, A. (2019, March). Identification of remote IoT users using sensor data analytics. In *Future of Information and Communication Conference,* pp. 328–337. Springer, Cham.

[32] Douligeris, C., and Mitrokotsa, A. J. C. N. (2004). DDoS attacks and defense mechanisms: classification and state-of-the-art. *Computer Networks*, 44(5), 643–666.

[33] Shahid, H., Ashraf, H., Javed, H., Humayun, M., Jhanjhi, N. Z., and AlZain, M. A. (2021). Energy optimised security against wormhole attack in IoT-based wireless sensor networks. *CMC-Computers Materials & Continua*, 68(2), 1966–1980.

[34] Zargar, S. T., et al. (2013). A survey of defense mechanisms against distributed denial of service (DDoS) flooding attacks. *IEEE Communications Surveys & Tutorials*,15(4), 2046–2069.

[35] Kaur, D., and Sachdeva, M. (2014). Impact analysis of ddos attacks on ftp services. In International Conference on Recent Trends in Information, Telecommunication and Computing.

[36] Draz, U., Ali, T., and Yasin, S. (2018, November). Cloud based watchman inlets for flood recovery system using wireless sensor and actor networks. In 2018 IEEE 21st International Multi-Topic Conference (INMIC), pp. 1–6. IEEE.

[37] Yasin, S., Ali, T., Draz, U., and Rasheed, A. (2018). Simulation-based battery life prediction technique in wireless sensor networks. *NFC IEFR Journal of Engineering and Scientific Research*, 6, 166–172.

[38] Humayun, M., Niazi, M., Jhanjhi, N. Z., Alshayeb, M., and Mahmood, S. (2020). Cyber security threats and vulnerabilities: Asystematic mapping study. *Arabian Journal for Science and Engineering,* 45(4), 3171–3189.

[39] Ali, T., Yasin, S., Draz, U., and Ayaz, M. (2019). Towards formal modeling of subnet based hotspot algorithm in wireless sensor networks. *Wireless Personal Communications*, 107(4), 1573–1606.

[40] Draz, U., Ali, T., Yasin, S., Waqas, U., and Rafiq, U. (2019, February). EADSA: Energy-aware distributed sink algorithm for hotspot problem in wireless sensor and actor networks. In 2019 International Conference on Engineering and Emerging Technologies (ICEET), pp. 1–6. IEEE.

[41] Draz, U., Ali, T., Yasin, S., and Waqas, U. (2018, December). Towards formal modeling of hotspot issue by watch-man nodes in wireless sensor and actor network. In 2018 International Conference on Frontiers of Information Technology (FIT), pp. 321–326. IEEE.

[42] Qaisar, Z. H., et al. (2020). Effective beamforming technique amid optimal value for wireless communication. *Electronics*, 9(11), 1869.

[43] Al-Musib, N. S., Al-Serhani, F. M., Humayun, M., and Jhanjhi, N. Z. (2021). Business email compromise (BEC) attacks. *Materials Today: Proceedings,* (2021).

6 Efficient Feature Grouping for IDS Using Clustering Algorithms in Detecting Known/ Unknown Attacks

Ravishanker, Monica Sood, Prikshat Angra,
Sahil Verma, Kavita, and NZ Jhanjhi

CONTENTS

6.1 INTRODUCTION

With the exponential increase in the growth of networks and diverse devices, it becomes a very complex task to manage network and its traffic flow. Due to the introduction of new protocols and newer technologies, a newer mechanism with real-time monitoring and controlling abilities is required to cope with these newer high-speed networking devices and networks. For smaller networks, similar aspects are also required since managing the devices now becomes very challenging in the context of a dynamic network topology and the trust within the network. A machine-learning approach is taking place to learn automatically from the previous network analysis. Recent approaches include using the blockchain method so that peer-to-peer nodes can be authenticated [1] and same can be applied to blockchain distributed network too. Blockchain can be applied in detection of intrusion detection also for generating an alarm for building an enhanced trust model [2].

The most common way to perform traffic monitoring is to look at each packet and impose a minimum and maximum threshold, or inspect each packet. If the current value is within the bounds, then the traffic is normal, and beyond this is abnormal, so an alarm would be generated. During the procedure of implementing the alarm, there may be the chances of sensing a large number of false alarms; that

DOI: 10.1201/9780367808228-6

is, an alarm may be generated even if the traffic limit or packet contents was from the trusted sources. There may be some condition also where an alarm may not be triggered and the attack on the system actually is done, but no alarm occurs. Consideration of the first problem wastes time and effort, as well as availability of the system. The second problem is more critical as it tends to real damage with the associated network recovery cost, and leads to resource availability. These types of systems are designed with predefined sets of rules and cannot identify newer attacks.

6.2 RELATED WORK

Every day, the cyberattacks become more engineered with high deal of complexity. The main challenge becomes deploying techniques that can recognize unknown attacks [3]; that is, those for which a predefined set of signature patterns are set for newer attack. Another issue is that data is coming from multiple sources, so it needs to handle and collect huge amounts of data from multiple data sources, and there is a chance that the collected data may not be relevant for detecting the attack. Multiple general-purpose algorithms have been designed for intrusion detection, i.e., misuse detection and anomaly detection. Machine learning can handle large data sets and identify and capture hidden patterns in a data set; this describes the grouping of data in a cluster using classification and regression to detect future prediction and using the resultant in a rule-based problem. Following security attacks can be identified with the help of machine learning in networks: denial-of-service (DoS), user-to-root (U2R), root-to-local (R2L), or probing. For predicting new attacks, the regression approach can be used.

For better results, machine learning needs training of data and a better clustering approach with specific sets of data. G. Chandrashekar et al. [4] showed the benefits of a feature grouping algorithm and suggested the design criteria, including simplicity, stability, number of reduced features, classification accuracy, storage, and computational requirements. The feature-grouping approach greatly enhances attack prediction by deducing dimensionality in data and identifying discriminating features that reduce computational overhead and increase accuracy [5,6].

[7] Mohammadi et al., in this paper, the author talked about intrusion-detection systems based on feature grouping/feature selection. Currently, every industry depends on internet and distributed denial of services and bots attacks become huge problems for these industries. In this paper, authors used filter and wrapper methods and different algorithms. They used linear correlation coefficients and cuttlefish algorithms for grouping/feature selection. They also showed that newly developed security is not capable or not fully compatible with the customer requirements, i.e., firewalls, user authentication, and data encryption give the breaches in security so new a security technique is identified, called the intrusion-detection technique. They also compromised with speed and accuracy, and they discussed that in today's scenario, feature grouping with latest techniques will produce better results and make networks secure. Here, filter and wrapper methods are named as feature grouping; they are based on different algorithms. Basically, there are three methods which are used in feature grouping, i.e., filter, wrapper, combination. In the filter method, we select the feature based on ranking or best performance. On the wrapper method, we select the feature based on searching;

it selects the set of features and performs a function that gives accurate results. In the combination, we combined both filter and wrapper methods and then found out best feature grouping and better results, which reduces the false positive alert rate. They also identified and made clear that no more null or fake alarms are generated due to null values of features or misconfiguration of features. So, there is a need to apply some rules on it by machine learning to solve these kinds of problems.

Feature selection includes three main methods. First is filter methods, which provide ranking according to importance, and they solve the most important aspect of computation cost [3,8]. The second one is wrapper methods [8–11], and third is combination methods [10,11]. They proposed a new method of feature selection, including categorization, filtering, and classification to improve the classification. The proposed method was examined on the KDD Cup 99 using both binary and multiple-class classification. The result of their method was fast, but it was giving more false alarms. These can be also implemented in security for IoT based devices for detection of attack [12].

Khalil El-Khatib [8] had used best features for detecting intrusions in WLAN in which they used filter and wrapper models for taking features by deducing the 38 features to only 8, and the results show that the learning time of the classifiers is reduced to 33% and accuracy increased by 15%. Their work gives the path for further work on identifying the performance of classifiers-based ANNs, as well as SVMs, MARSs, and LGPs. Jingping Song et al. [9] showed improved results with fuzzy C-means algorithmic rule to compose groups with feature selection reduced from 41 to 10. As we all know, network operators and many organizations rely on internet, so intrusion detection is very important. But traditional detection methods are specialized to detect previous attacks, but nowadays, attacks are unable to be detected by previous methods; so, techniques or features to detect them need to be improved. In this paper, the author proposed a feature-grouping method for intrusion detection. The author projected how to compose groups by calculating mutual information of each feature, how to rank them, and how to find the number of groups. To rank features, author uses fuzzy C algorithm to get G group. Fuzzy C algorithm divides the rank vector. Mutual information of each feature in a group is computed in this algorithm, labeling the class and getting maximum one. Initially, this technique is used to determine the amount of data in communication media. The performance of this technique is comparatively better and saves computational power.

The author concluded that this grouping is based on mutual information. By calculating mutual information of one feature and then comparing it with other features, relations between them are represented. To choose one feature in a group, mutual information and class labels are used. Experimental analysis is performed on knowledge discovery in databases (KDD) 99 data set; the author's proposed technique performed better than present algorithms. So, by using this technique, new algorithms can be generated; also, further investigation can be performed by using this technique on other data sets. In future, more effective and efficient methods can be proposed than mutual information theory. Using that kind of algorithm and technique, they also predict the future attacks also occur on live or network system. Similar types of feature grouping with machine learning approaches, along with attack-detection accuracy, are shown in Table 6.1.

TABLE 6.1

Comparison of Different ML Techniques With Different Feature Grouping Extraction and Data Set [13–15,17–19,20]

Ref	ML Technique	Dataset	Features	Results
Pfahringer [13]	Supervised Ensemble of C5 DTs (offline)	KDD Cup	all 41 features	DR Normal: 99.5% DR Probe: 83.3% DR DoS: 97.1% DR U2R: 13.2% DR R2L: 8.4% Training: 24 h
Pan et al. [14]	Supervised NN and C4.5 DT (offline)	KDD Cup	all 41 features	DR Normal: 99.5% DR DoS: 97.3% DR Probe (Satan): 95.3% DR Probe (Portsweep): 94.9% DR U2R: 72.7% DR R2L: 100% ADR: 93.28% FP: 0.2%
Moradi et al. [15]	Supervised NN (offline)	KDD Cup	35 features	2 Layers MLP DR: 80% ESVM DR: 90% 2 Layers ESVM DR: 87%
Chebrolu et al. [16]	Supervised BN and CART (offline)	KDD Cup	Feature Selection using Markov Blanket and Gini rule	DR Normal: 100% DR Probe: 100% DR DoS: 100% DR U2R: 84% DR R2L: 99.47%
Amor et al. [17]	Supervised NB (offline)	KDD Cup	all 41 features	DR Normal: 97.68% PCC DoS: 96.65% PCC R2L: 8.66% PCC U2R: 11.84% PCC Probing: 88.33%
Stein et al. [18]	Supervised C4.5 DT (offline)	KDD Cup	GA-based feature selection	Error rate DoS: 2.22% Error rate Probe: 1.67% Error rate R2L: 19.9% Error rate U2R: 0.1%
Paddabachigari et al. [19]	Supervised Ensemble of SVM, DT, and SVM-DT Offline	KDD Cup	all 41 features	DR Normal: 99.7% DR Probe: 100% DR DoS: 99.92% DR U2R: 68% DR R2L: 97.16%
Sangkatsanee et al. [20]	Supervised C4.5 DT (online)	Normal: Reliability Lab Data 2009 (RLD09) Attack:	TCP, UPD, and ICMP header Fields	DR Normal: 99.43% DR DoS: 99.17% DR Probe: 98.73%
Li et al. [21]	Supervised TCM K-NN (Offline)	KDD Cup	all 41 features 8 features selected using Chi-square	41 features: TP 99.7% 41 features: FP 0% 8 features: TP 99.6% 8 features: FP 0.1%

Chebrolu et al. [16] The main problem the author is discussing in this paper is that some of the features contribute less and are redundant to the detection process. So, we need to build intrusion-detection systems (IDS), which are effective and efficient in the detection of real-world systems. The author performed tests on selection algorithm. The author proposes hybrid architecture for real-world intrusion detection by combining different selection algorithms. These are Bayesian network (BN) and classification and regression trees (CART). Bayesian network (BN) selection algorithms perform under uncertain conditions. Bayesian networks work on (DAG) directed acyclic graph principle and calculate conditional probability of one node and allocate to another node. Classification and regression trees (CART) use binary recursive partitioning. Parent nodes always divide into two nodes; that's why process is binary and recursive because every node is treated as a parent, so the process repeats itself. In this paper, they investigated on different data-mining techniques; here, they select a feature or set of features for network data. They also observed the results by combining different features and different sets of features. They made an effective technique that detected the attack and generated a true positive alarm using above selection algorithm.

Author concluded the performance of new techniques in this paper for intrusion detection on the basis of DARPA benchmark. They proposed the combination of both Bayesian network (BN) and classification and regression trees (CART) classifiers by joining proposed hybrid architecture for intrusion detection. So, the result of analysis is very accurate. This technique performs better for probe, normal, and denial of service (DOS). The author's goal is to build more efficient and accurate classifiers for user-to-root (U2R) types of attacks, too. The author also makes subcategories of attacks and also gives a feature; mainly it also analyzes services that are hosting and leading in helping the attacks.

Davis et al. [22], in this paper, authors addressed the network traffic feature selection problem. They reduced the computational effort required to generate for live traffic observation for various network-based attacks. They also reviewed the data preprocessing for anomaly-based network-intrusion detection. To resolve above problems, the author recommended data-preprocessing techniques. This technique is a combination of analyzing and selecting the methods that have been used to detect intrusion detection. Both are coupled together to perform best searching and efficient efforts. Here, the author tried to find packet header features. Anomaly-detection techniques were used by network-intrusion detection system (NIDS) software and used together with machine-learning techniques, with particular use of models to detect features and compare both supervised and unsupervised algorithms. Making a group of features gives better understanding and machine-learning help to perform the model with better prediction.

By using this technique, the author concluded that with the grouping of features, the strong contribution can identify the relevant parts of network or live traffic with better accuracy prediction. It leads to deep packet inspection and thus reduces computational effort for online feature generation. Nowadays, security becomes a challenge for each and every organization and department. To solve that kind of problem, we need anomaly detection in network traffic/live traffic with proper feature knowledge.

Ma, C., [23] Nowadays, security becomes a challenge for each and every organization and department. To solve this problem, anomaly detection in network traffic is needed. So, to have accurate and effective traffic features for anomaly detection is big challenge. In this paper, the author proposed methods that show relatively better performance than present algorithms. The method proposed by the author for network anomaly detection comprises analysis of environmental, sequential, and statistical characteristics. Their proposed method is a combination of hybrid neural networks; these are one-dimensional convolutional neural network model (1D-CNN) and dynamic neural network (DNN) schemes known as

(AD-H1CD), anomaly detection for network flows based on hybrid neural network. This method is comparatively better and its analysis all above stated characteristics.

In this paper, the proposed method is highly accurate and efficient when compared to previous network anomaly detection algorithms. Anomaly detection for network flows based on hybrid neural network (AD-H1CD) extracts multistage features of network flow and analyses them. The detection rate of this algorithm in multi-arrangement has been improved. Every time they compared new techniques with old, and made a change on new thinking and doing efforts to make it more efficient and effective as per needs in new technology and future needs. It also gives better outcomes, and the predication rate and accuracy rate is high if it is considered with the old and new techniques to detect attacks.

[24] Sheikhan et al., to secure the communication system, intrusion-detection is needed. So, the author proposed a new detection technique to improve rates of detection. But the proposed technique is also not efficient; false alarm rates are still not degraded in this technique also, but its detection rate is comparatively better. The author proposed three-layer recurrent neural network architecture. Recurrent neural network (RNN) partially connected with hidden layers is used as misuse-based intrusion-detection system (IDS). Recurrent neural network (RNN) input is categorized as features, and recurrent neural network (RNN) output is projected as misuse-based intrusion-detection system. Due to its reduced size, its detection rate is comparatively better. Categorization of input features are content features, basic features, host-based traffic features, and time-based traffic features. Types of attacks are DoS (denial of service), R2L (remote-to-local), U2R (user-to-root), and probe. Recurrent neural network (RNN) basically composed of five output neurons, i.e., it's four types of attacks and represents normal classes.

Author concluded that new technique recurrent neural network (RNN) performed more accurately than previous ones. Recurrent neural network (RNN) are composed of partially connected layers with four features as input and misuse-based intrusion detection system (IDS) as output. Results show that the reduced size improved the analysis performance. It shows good results on remote-2-local (r2l) attack comparatively to other classifiers. Multiple attacks are identified here, and categories also, with the help of group featuring. They made a group of features and indicated with the features that if these features are combined, then the possibility of attacks is very high. They also perform analysis on each and every feature that gives better and good results and also helps in making groups as per changes. They also made a different combination for future analysis, and if small changes in features occur, then the tendency may lead to some type of network attack mostly.

[25] Ponkarthika et al., communication and information technologies are increasing day by day; nowadays, information is shared at high rates all over the world. So, in this growing era of information sharing, cyber threats or attacks are a major issue. Intrusion detection systems detect the network security breaches, which are helpful for detecting attacks, but the main challenge is the efficient and accurate performance of the network-intrusion detection system (NIDS). There is need for an excellent network-intrusion detection system to detect unpredictable attacks and security breaches, too. To implement flexible and efficient network-intrusion detection systems, the author proposed a deep learning-based model. Deep

learning uses complex architecture to achieve high data abstraction. To achieve high detection rates, we use (LSTM) long short-term memory in (RNN) recurrent neural network and apply it on an intrusion-detection system (IDS) model using knowledge discovery in databases (KDD) cup 1999 data set. The author analyzed the performance of this approach on knowledge discovery in databases (KDD) cup 1999 data set. So, this is an effective approach to train sequence of data. Previously proposed recurrent neural network (RNN) has many problems, so to resolve them, long short-term memory (LSTM) architecture is used. Basically, using grouping for the selection method and using mutual information by assessing performance of existing techniques may improve.

The author concluded that the main goal of this method is to improve the intrusion-detection system (IDS) from the past observations. He implemented a new intrusion-detection system (IDS) classifier using long short-term memory (LSTM) architecture in recurrent neural network. By comparing the results of this method with other classifiers, the author concluded that attacks are well detected with this classifier. So, in future, we can further investigate and improve the performance of intrusion-detection systems (IDS) from this proposed intrusion-detection system (IDS) model, too. Feature selection is also useful when we reduce computational time and in understanding data. Feature selection is also used in optimized classifiers by removing irrelevant features from existing ones.

[25] Mohamad et al., here, the author has proposed a machine-learning method that uses a hybrid learning technique for intrusion detection on a network by the combination of classification such as support vector machine (SVM) and k-means clustering. It detects threats; the main threat is intrusion; here we make a technique or feature collection that helps to transmit data over a network. As author observed here that if they use machine learning and k-means clustering, along with support vector machine (SVM), then the outcome is improving the performance. Various techniques, methods, and approaches are developed to negotiate with limitations of intrusion-detection systems and also reduce the false alarm, time, and low accuracy. Main aim is to focus on improving the detection rate and improving the false-negative alarm, and reducing the false negative alarm.

Performing a hybrid machine-learning technique, the results show a new achievement here: they got the positive detection and also reduced the false alarm rate. Here, it successfully identified an attack occurring on the network and detected an alarm; the most important part is the accuracy is also high. Mainly it is done by two machine-learning techniques, i.e., support vector machine (SVM) and k-means clustering. Here, author purposed a system that helped to understand a mechanism that included features. With the help of different techniques and algorithms, it gives a clear review or comparison between good or bad features also.

Maralani et al. [26], in this paper, the author is using a hybrid method of support vector machine (SVM) and a genetic algorithm (GA) is proposed; then an intrusion-detection system (IDS) was implemented that can explain the problem. Here, the author also finds the solution for multi-subsets and for feature-set grouping also. Genetic algorithms (GA) are used to solve a problem with proper set for feature grouping. In this technique, the author applied an artificial intelligence (AI) technique to solve the problem with the help of a support vector machine (SVM) and a genetic algorithm (GA). An artificial intelligence (AI) technique proved that it can solve that problem that is nonlinear and nonconvex. Author mainly focused their techniques on support vector machine (SVM) and genetic algorithm (GA). It gives very good solutions and provides better interface and results for feature grouping or sets. Additional parameters of features are also considered. They also measure here some machine-learning methods that include fuzzy logic-nearest neighbor, support vector machine (SVM), artificial neural network (ANN), artificial immune system (AIM), and genetic algorithm (GA) for comparison purpose.

In this paper, author used hybrid methods of support vector machine (SVM) and genetic algorithm (GA) to reduce the number of features utilized in a proper manner that makes it and fits it into a priority wise manner. Author makes categories of that and fit features on that according to their priority. They proposed an approach that gives better true positive results, which helps in detection of attacks occurring in the network. It also reduces the false positive rate. It also differentiated between the attacks with the help of features. Here, they mainly focused on security and the detection; we also understand here the importance of sensitive data. We analyze multiple results here out by applying different techniques and algorithms.

[27] Iglesias et al., in this paper, the author addresses the network traffic feature-selection problem. He reduces the computational effort required to generate for live traffic observations. The old traffic selection feature contains redundancy among features, irrelevant features, and interdependencies among features. To resolve above problems, the author recommends multistage selection feature technique with filters and step regression wrappers. This technique is a combination of wrappers and filter techniques. Both are coupled together as wrappers, and filters perform best searching and lower the computational efforts. In stepwise regression, we choose the suitable feature, and it is based on greedy algorithms. Hence, in this they used a forward selection and backward elimination wrappers technique. Better understanding of features was proposed by the author, as well as better analysis and findings based on most-used features, which show high changes in the environment were taken.

By using this technique, author concluded that features which display strong contribution are 16, those having low contribution are 14 and 11 features have zero contribution. Initially, author performed testing on five fundamental classifiers and observed negligible difference in performance. He then analyzed the cost for generating features and eliminated 13 costly features. Thus, computation effort was reduced for online feature generation. So, author experiment analyzed the irrelevant features. Here, they get correlation and finding the most changing features in the total features. They also categorized that and used that in an effective manner. Making a group of features gives better understanding of attacks, and it also represented it and detected it in an easy way and also indicates an attack occurring on the network. Basically, it also increases the rate of true positive and reduces the rate of false positive.

6.3 DESIGN METHODOLOGY

Feature selection is an important aspect for attack categorization. Various attacks on which feature selection is based are denial-of-service (DoS), user-to-root (U2R), root-to-local (R2L), or probing. Feature-selection criteria should be based on matching the attack criteria. If grouping is done with more relevant features, then the attack will be more accurate to identify. In KDD-CUP99, there were 41 features available; they can be further categorized based on traffic categories like DoS, U2R, R2L, probe, and normal. Out of KDD-CUP99, not all features are relevant. Some may not be useful in detecting attacks, while some of the features may not be useful for classification.

Algorithm:

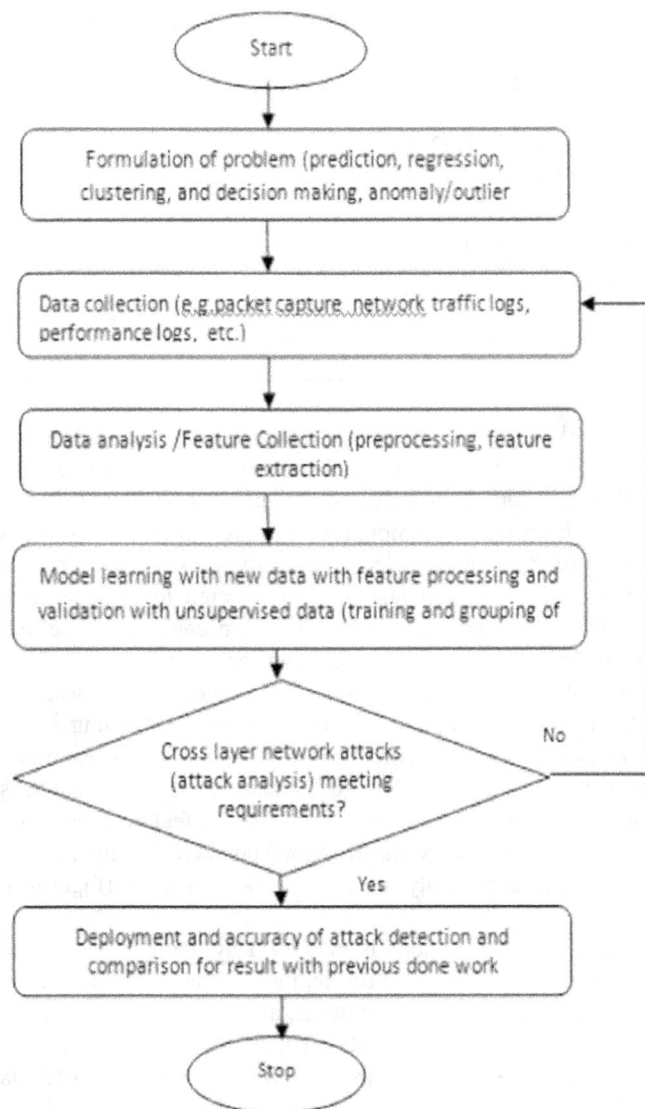

Step 1. Prepare network for data-processing phase.
Step 2. Capture packet and network traffic logs from interface i in real-time using snort or wire shark or tcpdump.
Step 3. Extract IP features F.
Step 4. Create feature selection group G.
Step 5. Classify data for training data t1 and t2.
Step 6. Perform training on data sets and using adaptive learning to apply classifier techniques for further testing.

Let T -> t1 and t2 as labeled and unlabeled data from feature grouping, N for null data, // some packet may contain incomplete features
Start the training set as t1;
While (t2 is not null)
{
select one instance i from t2 and generate alarm;
// FP, FN, TP, TN
If (C (i) <null)
Add i to t1 and remove i from t2;
}
Output classier of t1.
Step 7. Save result in logs file and generate alarm.

When evaluating a network, it passes a huge data and generates logs, so for small networks also, these data are very large for an IDS. Network traffic contains complex features, and classification of these features is difficult to relate with each other. First task is to do data filtering so that not all data are required to be processed by an ID. Second task is to do feature selection where relevant features must be grouped together so that only relevant features are selected to same time and computing complexity. The third aspect is clustering approach to relate these features together. Some of the techniques involved Bayesian networks, CART, etc. Finally, ranking of these features is required using SVM or neural networks.

Feature selection helps in reducing the classification process for training of data. Also, feature selection needs to consider according to the IDS type. For example, a behavior-based IDS and a network-based IDS need to consider different feature selection and feature grouping. For classification purposes, a set of feature F need to be taken for target output T. This is followed by classification approach. Taking a minimal feature will reduce timing and complexity in classification process (Figures 6.1 and 6.2).

Results clearly show that the performance is increased by using fewer features compared with using all 41 features. In normal traffic, the accuracy level is 99.57 by taking all 41 features, and when using only 17 features, accuracy is 99.64. For other attack classes, since fewer features were selected, the accuracy level is decreased a bit. Experimental results from previous research shows how experimental data sets using different machine-learning approaches affect results in detecting known attack (Figure 6.3 and Table 6.2).

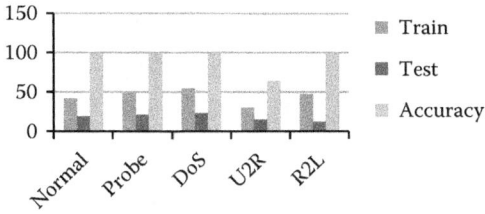

FIGURE 6.1   Performance by using 41 features.

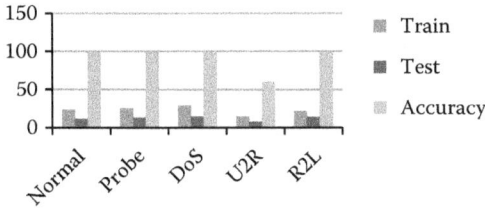

FIGURE 6.2   Performance by using 17 features.

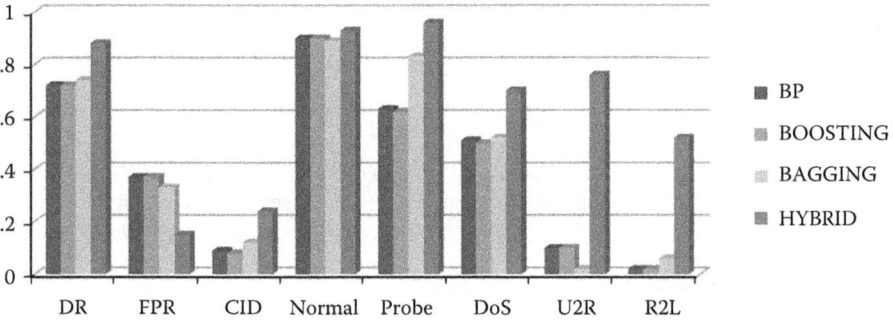

FIGURE 6.3   Experimental result showing performance analysis using various ML approach [22].

TABLE 6.2
Attack Class, Training and Test Record of KDD Data Sets [22]

Type	Training Record	Test Record
Normal	100	500
Probing	100	75
Denial_of_service	100	75
Root_to_local	100	50
User_to_root	11	50

6.4 APPLICATIONS AND FUTURE WORK

All the past work done on feature grouping using clustering and combination of AI approach can be extended to combine with blockchain techniques to create trust among various dynamic interconnected nodes; that way, when they join the network, it can be monitored on a real-time basis and a malicious node can be identified easily to thwart any malicious task. If the source and destination are bound with blockchain, then it would be easy to identify attacks at application layers. This could be specially done in E-health sector where medical data is shared with patients and the concerned doctors. The machine could be set up with a combination of user applications containing block chain and IDS so cross-layer attacks can also be detected.

6.5 CONCLUSION

Feature selection and grouping methods need various factors such as type of IDS, attack types, preparing data for testing, and training purpose. Not all features are useful in the attack-detection process due to redundancy, and some features may not be useful due to lack of its participation in detecting attack. Feature selection helps in enhancing the performance of IDS with respect to its complexity and cost. In future work, blockchain implementation on the application layer is examined for application and the network layer attack in combination with network based intrusion-detection system.

REFERENCES

[1] Alexopoulos, N., Vasilomanolakis, E., Ivanko, N. R., and Muhlhauser, M. (2017). Towards blockchain-based collaborative intrusion detection systems. In *Proceedings of the 12th International Conference on Critical Information Infrastructures Security*, pp. 1–12.
[2] Kok, S. H., Abdullah, A., Jhanjhi, N. Z., and Supramaniam, M. (2019). A review of intrusion detection system using machine learning approach. *International Journal of Engineering Research and Technology, 12*(1), 8–15.
[3] Vijayalakshmi, B., Ramar, K., Jhanjhi, N. Z., Verma, S., Kaliappan, M., Vijayalakshmi, K., Kavita, Ghosh, U., and Vimal, S. (2020). An attention based deep learning model for traffic flow prediction using spatio temporal features towards sustainable smart city. *International Journal of Communication Systems, 34*(3).
[4] Chandrashekar, G., and Sahin, F. (2014). A survey on feature selection methods. *Computers & Electrical Engineering, 40*(1), 16–28.
[5] Batra, I., Verma, S., Kavita, and Alazab, M. (2019). A lightweight IoT based security framework for inventory automation using wireless sensor network. *International Journal of Communication Systems, 33*(2), e4228.
[6] Rani, P., Kavita, Verma, S., and Nguyena, N. G. (2020). Mitigation of black hole and gray hole attack using swarm inspired algorithm with artificial neural network. *IEEE Access, 8*.
[7] Humayun, M., Niazi, M., Jhanjhi, N. Z., Alshayeb, M., and Mahmood, S. (2020). Cyber security threats and vulnerabilities: A systematic mapping study. *Arabian Journal for Science and Engineering, 45*(4), 3171–3189.

[8] El-Khatib, K. (2010). Impact of feature reduction on the efficiency of wireless intrusion detection systems. *IEEE Transactions on Parallel and Distributed Systems*, *21*(8), 1143–1149.

[9] Elijah, A. V., Abdullah, A., Jhanjhi, N., Supramaniam, M., and Abdullateef, B. (2019). Ensemble and deep-learning methods for two-class and multi-attack anomaly intrusion detection: An empirical study. *International Journal of Advanced Computer Science and Applications, 10*, 520–528.

[10] El-Alfy, E. S. M., and Alshammari, M. A. (2016). Towards scalable rough set based attribute subset selection for intrusion detection using parallel genetic algorithm in MapReduce. *Simulation Modelling Practice and Theory, 64*, 18–29.

[11] Hussain, K., Hussain, S. J., Jhanjhi, N. Z., and Humayun, M. (2019). SYN flood attack detection based on Bayes Estimator (SFADBE) for MANET. In *2019 International Conference on Computer and Information Sciences (ICCIS)*, pp. 1–4. IEEE.

[12] Batra, I., Verma, S., Kavita, Ghosh, U., Rodrigues, Joel J. P. C., Nguyen, G. N., Sanwar Hosen, A.S.M., and Mariappan, V. (2020). Hybrid logical security framework for privacy preservation in the green Internet of Things. *MDPI-Sustainability*, *12*(14), 5542.

[13] Pfahringer, B. (2000). Winning the KDD99 classification cup: Bagged boosting. *SIGKDD Explorations, 1*(2), 65–66.

[14] Pan, Z. S., Chen, S. C., Hu, G., and Zhang, D. Q. (2003). Hybrid neural network and C4. 5 for misuse detection. *Training, 4*(5).

[15] Moradi, M., and Zulkernine, M. (2004, November). A neural network based system for intrusion detection and classification of attacks. In *the IEEE International Conference on Advances in Intelligent Systems-Theory and Applications*, pp. 15–18.

[16] Chebrolu, S., Abraham, A., and Thomas, J. P. (2005). Feature deduction and ensemble design of intrusion detection systems. *Computers & Security, 24*(4), 295–307.

[17] Shahid, Hafsa, Ashraf, H., Javed, H., Humayun, M., Jhanjhi, N. Z., and AlZain, M. A. (2021). Energy optimised security against wormhole attack in IoT-based wireless sensor networks. *CMC-Computers Materials & Continua, 68*(2), 1966–1980.

[18] Stein, G., Chen, B., Wu, A. S., and Hua, K. A. (2005, March). Decision tree classifier for network intrusion detection with GA-based feature selection. In *the 43rd Annual Southeast Regional Conference-Volume 2*, pp. 136–141. ACM.

[19] Peddabachigari, S., Abraham, A., Grosan, C., and Thomas, J. (2007). Modeling intrusion detection system using hybrid intelligent systems. *Journal of Network and Computer Applications, 30*(1), 114–132.

[20] Li, Y., and Guo, L. (2007). An active learning based TCM-KNN algorithm for supervised network intrusion detection. *Computers & Security, 26*(7–8), 459–467.

[21] Kumar, G. (2019). An improved ensemble approach for effective intrusion detection. *The Journal of Supercomputing*, 1–17.

[22] Davis, J. J., and Clark, A. J. (2011). Data preprocessing for anomaly based network intrusion detection: A review. *Computers & Security, 30*(6–7), 353–375.

[23] Jhanjhi, N. Z., Brohi, S. N., Malik, N. A., and Humayun, M. (2020). Proposing a hybrid RPL protocol for rank and wormhole attack mitigation using machine learning. In *2020 2nd International Conference on Computer and Information Sciences (ICCIS)*, pp. 1–6. IEEE.

[24] Almrezeq, N. (2021). Exploratory study to measure awareness of cybercrime in Saudi Arabia. *Turkish Journal of Computer and Mathematics Education (TURCOMAT), 12*(10), 2992–2999.

[25] Mohamad Tahir, H., Hasan, W., Md Said, A., Zakaria, N. H., Katuk, N., Kabir, N. F., and Yahya, N. I. (2015). Hybrid machine learning technique for intrusion detection system. In *International Conference on Computing and Informatics*.

[26] Aslahi-Shahri, B. M., Rahmani, R., Chizari, M., Maralani, A., Eslami, M., Golkar, M. J., and Ebrahimi, A. (2016). A hybrid method consisting of GA and SVM for intrusion detection system. *Neural Computing and Applications*, 27(6), 1669–1676.

[27] Iglesias, F., and Zseby, T. (2015). Analysis of network traffic features for anomaly detection. *Machine Learning*, 101(1), 59–84.

7 PDF Malware Classifiers – A Survey, Future Directions and Recommended Methodology

N.S. Vishnu, Sripada Manasa Lakshmi, Kavita, Sahil Verma, and Awadhesh Kumar Shukla

CONTENTS

DOI: 10.1201/9780367808228-7

117

7.1   INTRODUCTION

Current attackers are honing their deception techniques for developing stable malware using obfuscation techniques to avoid detection and boost their destruction capabilities. Much of the malware attacks are conducted with frequently used files, such as photos, PDFs, spreadsheets, and other files by introducing the malignant code. An intruder needs to have information about the structure of certain benevolent files in order to merge dangerous code with benevolent files. The embedding of the malware with these routinely used documents has three key advantages. The first advantage is that the majority of people are relying on PDF files, which could increase the chances of social engineering attacks on people [1]. The second advantage is that a large number of users are using obsolete technologies and applications, making them vulnerable to such attacks [2]. Vulnerabilities in software can be exploited to construct successful malware that prevents detection. The more the victim chooses to use obsolete and insecure software, the longer the lifetime of their systems' malware infection [3]. Many researchers showed that self-learning strategies can be used to boost the detection skills and performance of the detection systems. For example, consider the system to be a child and the data which is imparted to the system as the chapters taught to the child at school. The main aim of the attacker would destroy the system, while in our example case, to fail the child in the assessment. So, to achieve this outcome, the attacker may either try to feed the system with illegitimate data or give carefully designed input to confound the system. In the case of the example, the lecturer may try to teach the child with inappropriate data or even give complex questions at the time of exam to fail the child. The feeding of inappropriate data can be considered as a poisoning attack, and the process of rendering a tricky question during the assessment can be called an evasion attack. The majority of research administered on the self-learning detection techniques have manifested that these methodologies are capable of precisely recognizing and classifying the files blended with hidden malware [2]. The research work conducted on carrying out evasion attacks on this kind of detection system has staged that a diligently fabricated input to these systems can conquer them [2]. For example, adding perturbations to a clear image was able to misclassify it by a machine-learning image recognition system.

In this paper, we will be discussing the different machine-learning-based PDF classifiers, which were proposed by different authors. The portable document format or PDF files are most widely used by most users due to their being lightweight and having high portability. It is the widely inherited format for viewing and sharing the data, which makes it the hacker's choice of vector for infection. Here we will be discussing the PDF structure and different methodologies by which the malware PDF-detection systems can be evaded by creating the adversarial samples.

These PDF detectors are mainly of two different types, static and dynamic. We have reviewed most of the PDF classifiers proposed through the present day for grasping the mechanisms these detectors use for detection.

7.2   X-RAY OF PDF FILE

The way of parsing a PDF can depend upon the type of parser used to manipulate or view it [2–6]. There are few basic elements in every PDF file that are supreme for the representation of PDF files [2]. These elements are basically classified in to two categories, the general structure and file content [2]. The first category includes the information regarding the content of storage within the file [2]. The second category describes the way the contents are displayed to the user accessing the file [3]. The general structure of a PDF file can be considered as a network of objects in which each of these objects will have distinct functions to perform [2]. There are four main components in the PDF structure; they are header, body, cross-reference, table and trailer. The header is the piece of the PDF that comprises one-line text code [7]. This content is started with a % sign and implies the version of the PDF document. Body part comprises a request for objects, which means the exercises done by the PDF document. Inside these items, there are odds of including various types of mixed document types like pictures and video records [2]. Indeed, even executable vindictive codes can be embedded into PDF records by presenting them in at least one object of the body part [2]. Each article proclaimed in the body is a piece of PDF record containing an unmistakable number, which is its reference number [2]. The reference number can be helpful in calling a specific item. At whatever point an item should be known, the reference number can be used. Each unmistakable article, which is pronounced in the body, closes with a catch-phrase marker "endobj" [3]. By cross-reference table, we can understand that this piece of the PDF contains a table. We should comprehend the usefulness and job of this table [2]. This table by and large contains the data about the area of different articles inside the PDF document. It additionally advises the parsers from where to start the parsing measure [3]. It is addressed in the record utilizing "xref" markers, and under that, we can see different lines where each line demonstrates the references to various items referenced in the body field. Just those objects that are referenced in the cross-reference table will be parsed at the hour of parsing the PDF record [2]. Trailer field portrays a portion of the fundamental components that are vital for the record translation [2]. The address of the first item goes under one of them. It likewise contains the references to the metadata related to the references [3]. The end of the trailer field is indicated by utilizing a marker "%%EOF." This likewise means the end of the file and signals the parser to stop its parsing movement [2].

7.3   READING OF PDF FILES

At the point when a specific PDF document is chosen for parsing by a parser, it plays out this action in the following way: Upon starting, the parser goes to the trailer field and decides the principal object to be deciphered [2]. At that point, grouping of items is parsed with the assistance of references mentioned in the cross-reference table. At that point, at long last, the parsing methodology is ended when it identifies the "finish of file marker" [7]. One of the huge highlights of utilizing PDF documents is that they take out the prerequisite of reproducing records without any preparation when some new articles are brought into the previously existing records [2]. Considering all these things, they make new field structures that are devoted to putting away the recently presented substance [4].

7.4   STEPS INVOLVED IN CLASSIFYING PDF FILES

"AI" innovation is broadly being used practically in each field [1]. This innovation helps in making the framework wiser and empowers it for dynamic [1,8,9] use. Because of its wide scope of utilization, it is likewise conveyed in deciding malignant substance covered up in records [10]. Here we will talk about how this innovation can be aided successfully by identifying the threatening code in a PDF record under investigation [11,12]. A few self-learning frameworks were proposed by scientists for examining the PDF records in the previous decade [2,4,13,14]. In this part, we will examine the engineering of the self-learning-based malicious PDF finders and the means associated with ordering a given PDF record (Figure 7.1).

The fundamental goal of these frameworks is to productively order whether a particular document goes under the benevolent classification or the malignant one [15]. For confirming that, the framework needs to inspect and measure the examples and designs of the record's inner parts [10,15–19]. Despite the fact that there are numerous proposed identifiers, every one of them is seen to follow these three fundamental methods to examine the PDF records [8], which are:

7.4.1   PRIOR PROCEDURE BEFORE THE ACTUAL PARSING

This is the principal interaction completed on the PDF record that is picked for the review reason. In this technique, the inward code, which are basic, for example, JavaScript or ActionScript code, are executed in a steady climate under separation condition to comprehend its temperament [2]. It is prepared under detachment conditions with the goal that its execution does not influence the framework [20]. The other fundamental information components, for example, metadata of inward articles and watchwords are likewise removed from the PDF documents [11].

7.4.2   FEATURE EXTRACTION

This is the following movement completed by the detectors after the "pre-processing" action is finished. This technique uses the proffered information acquired from its former advance to achieve its examination action [3,11,21–27]. In this progression, the information extricated from the past stage is changed to the type of vector numbers that will actually want to pass on the presence of specific segments, for example, catchphrase and "programming interface calls" in the PDF document [28].

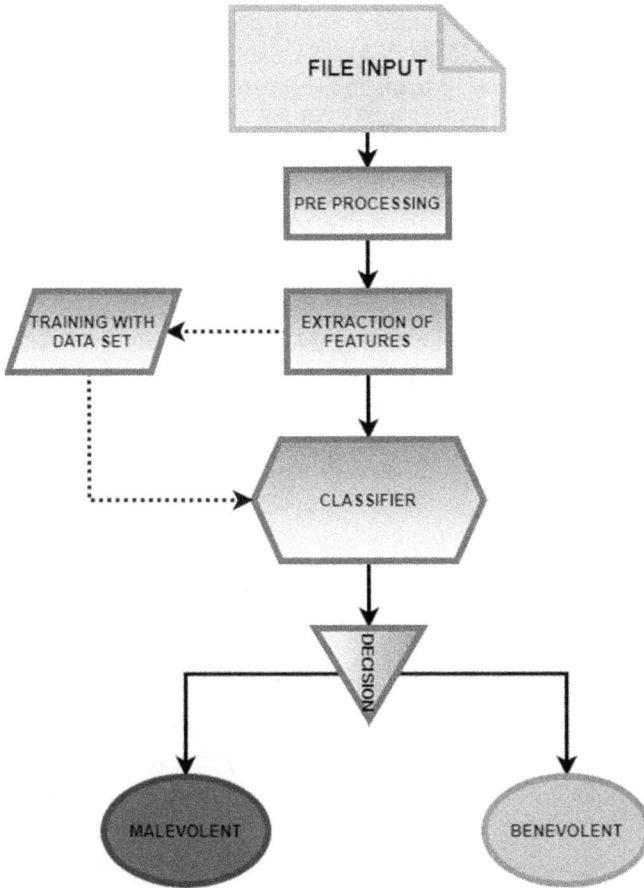

FIGURE 7.1 Work flow of PDF malware classifier.

7.4.3 CLASSIFIER

This is the stage where the algorithms are conveyed in dissecting the separated data from the past stages [16]. The classifiers are first prepared with test sets to build their information about deciding the contrasts between a generous and a threatening PDF record [29]. A framework can deliver quality effectiveness in arranging the PDF records, on the off chance that they are prepared productively [19]. The adequacy of the framework is not exclusively relied on the algorithm, yet in addition on the preparation set. The way toward preparing is led before the framework is really sent for the ongoing application [9]. The fundamental target of the classifier is to exactly order an offered document to be malicious or not [2].

7.5 MODES OF CLASSIFICATION

The malware indicators worked for examining the PDF records by helping self-learning strategies to be arranged in two primary classifications, static detectors and dynamic detectors [11]. The static identifiers perform static investigation of the record without executing the code inside the document [1,8–11,16–20,28–30]. The dynamic finders play out their examination by executing the genuine code in steady natural conditions [1,8,9,11,11,11,20,28–30]. We will examine more about these two recognition types beneath.

7.6 STATIC CLASSIFICATION OF PDF FILES

A large portion of the malware assaults actualized by using the PDF records utilize JavaScript or ActionScript code [11]. Thus, these kinds of identifiers primarily center around the areas of the document where these sorts of codes are recognized. In any case, we cannot infer that a specific PDF record is a vindictive one just by taking a gander at the presence of the JavaScript code. Indeed, even benign records may likewise have such codes in them [11]. Along these lines, the static identifiers investigate for the examples of utilization of explicit factors, watchwords, capacities, and programming interface calls for deciding the idea of the document [28]. PJScan is a static PDF locator that chose the malignance of the PDF records dependent on the recurrence of use of variables, administrators, and capacities [31]. A portion of different locators depended on tokenization and matching strategies [28]. In any case, in situations where these JavaScript substances are vigorously obscured by the assailant inside the PDF records, the static classifier may flop by misclassifying the record [1]. The assailant can make the noxious substance especially covered up, with the goal that it can't be distinguished by the parser at the hour of parsing cycle, and this code can be powerfully called at runtime to complete the threatening activities. Thus, the primary defect existing in this class of detectors is that they can't be prepared to precisely order a purposely obfuscated tests [2] (Figure 7.2).

7.7 STATIC CLASSIFICATION SYSTEMS

7.7.1 PJScan

Pavel Laskov and Nedim Šrndic [31], in 2011, proposed a static malware detection framework that was ready to distinguish the dangerous JavaScript content contained inside the document [31]. Creators guaranteed that their model had to have brought down the preparation time needed for examining the record compared to the recently introduced models, and it was likewise showing a perfect viability in recognizing known and obscure malevolent sorts [31]. This technique was most appropriate for preparing enormous data sets due to its low review time and high effectiveness [31] (Figure 7.3).

The architecture of the PJScan model appears in the figure. At the point when a specific PDF record is given to the framework for examination, the record goes through a component-extraction module that contains subcategorial modules, for example, a JavaScript extractor to draw out the JavaScript code inside the record substance [31]. At that point, all the passages of these JavaScript code are then appointed with tokens [31]. At that point, the calculation is utilized for the grouping reason.

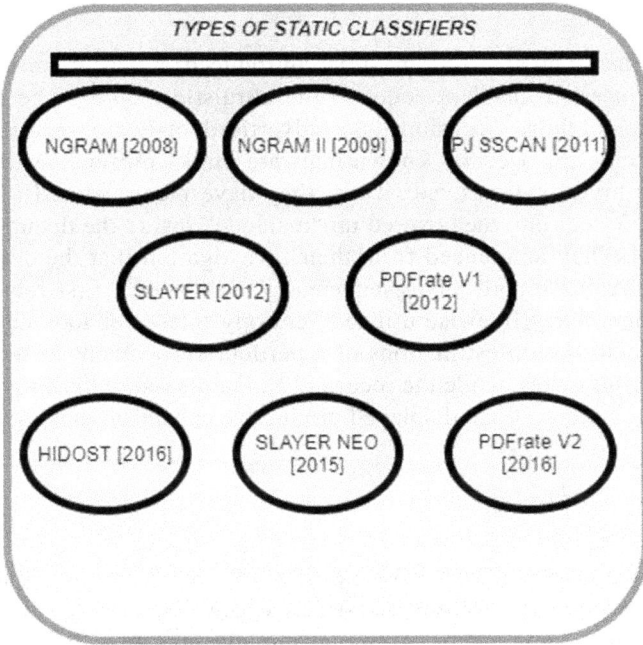

FIGURE 7.2   Static PDF classifiers.

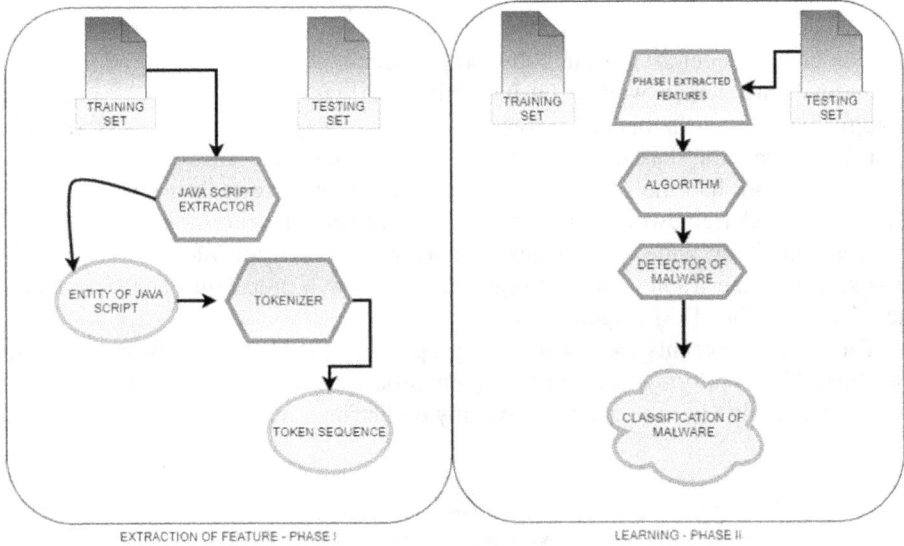

FIGURE 7.3   Different phase flow of PJScan [31].

7.7.2 N-Gram

M. Zubair Shafiq et al. [32], in 2008, introduced a way to distinguish the presence of inserted malware codes in the altruistic records. The authors expressed that, on those occasions, the advertised malware security software could not decide the recently known malware marks present inside the documents at the investigation cycle. Thus, they have manufactured a framework that was fit for deciding the covered up "malcode" inside the documents called N-Gram [32]. They announced from their investigation that the dominant part of the non-noxious record "byte sequencing" exhibited a first request dependence structure. They likewise utilized "entropy rates" for looking at the distinctions of distribution estimations of a particular document to recognize the home of harmful codes inside the record [32]. On assessing the above proposed model by the authors, it had displayed productive execution in managing mixed malicious records.

7.7.3 Slayer

Davide Maiorca et al. [17], in 2012, proposed a module to remove the features inside a PDF document under the assessment cycle. The authors have introduced a viable classifier and feature extraction module that were joined to shape a compeling static detection model. This device was said to proffer high flexibility [17]. One can use the instrument either as a "remain solitary apparatus" or as an extra "module highlight" to improve the ability of existing recognition frameworks [17].

7.7.4 N-Gram II

S. Momina Tabish et al. [33], in 2009, propounded a novel methodology for the ID of the malware, which worked by inspecting the bytes of the file. In this methodology, the substances that are analyzed are not put away for any future purposes; it is a "non-signature" based methodology. Subsequently, the creators asserted that their framework had the capacity to distinguish the obscure malware assortments. The introduced framework was considered in contrast to enormous examples of records containing wide assortments of malware families and file designs [33]. It was seen that the framework had organized to show a precision of 90% in characterizing the files [33] (Figure 7.4).

The figure represents the engineering proposed by S. Momina Tabish et al. [33] in 2009. The creators guarantee that their procedure used to construct this framework can wipe out the need to have any earlier information with respect to the

FIGURE 7.4 Flow of N-Gram II classifier [34].

kind of document picked for the investigation reason [33]. Consequently, this model might distinguish the malware in situations where the attackers attempt to control the header segment of the PDF records [33]. In the architecture introduced by the authors, the framework is separated into four unique modules [34]. The main module is the "square generator module" that partitions the "byte level content" inside the record into equivalent estimated and more modest squares [34]. The following module, "highlight extraction module" plays out the computational measurements on the squares produced in the past level [34]. The other two modules in particular, "arrangement and correlational" module do different examination on the results got from the past model to identify the indications of malware [34].

7.7.5 PDFRATE v1

Charles Smutz and Angelos Stavrou [16], in 2012, propounded an architecture for catching the vindictive attributes of PDF records in a robust manner by helping self-learning draws near. This technique was fit for capturing the data from the metadata of articles present in the PDF structure [16]. This locator worked on the premise of "irregular forest classification" technique, which is a classifier prepared to perceive the highlights from the different "order trees" shaped [16]. This classifier was well known to deliver great recognition rates, even in instances of obscure malware qualities [16]. The "irregular forest classification" technique is as yet utilized in a large number of the recently proposed location frameworks because of its proficient systems [16] (Figure 7.5).

The figure shows the course of action made by Charles Smutz and Angelos Stavrou [16] in 2012 for the proficient malware identification. This framework goes through two degrees of testing. The primary testing is directed to order the document into either benevolent or noxious class [16]. In the event that the record goes under the noxious class, at that point, the document is made to go for another degree of grouping level to decide, regardless of whether this vindictive document is a deft one or a focused one [16].

7.7.6 HIDOST

Nedim Šrndic and Pavel Laskov [29], in 2016, introduced a static detection framework called Hidost for effectively distinguishing the non-executable malware content present inside the PDF record structures. As the assailants started to annex non-executable code into the kind-hearted documents to dispense with the location by static, and also dynamic, classifiers, there came the need of developing frameworks to recognize these sorts of malware dwelling in the records [29]. This paper was an augmentation to the recently proposed approach by the equivalent creators in 2013 [29]. In that paper, the authors introduced a methodology of fusing the content with their sensible components present in the document for accomplishing a raised order exactness. Albeit this model was created to identify the malware in the PDF and Flash records, the authors state

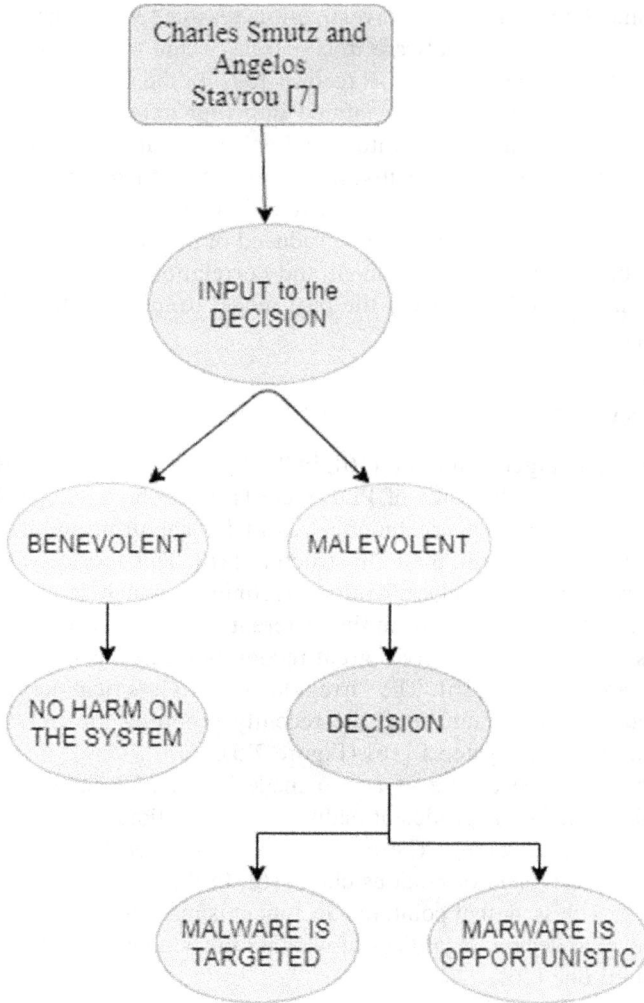

FIGURE 7.5 Workflow in PDFrate v1.

that their methodology can be stretched out to recognize malware in different arrangements. On assessing this framework, it was discovered that this methodology has ruled dominant part of the business VirusTotal location programming in precision [29].

The authors guarantee that their framework is fruitful in separating between the malicious and the favorable documents. We cannot accept just a single vector as a typical component for arranging various documents [9]. The handling and identification strategies are isolated into various strides in this model with the goal that this framework can be further stretched out to proficiently distinguish other document malwares [9].

7.7.7 SLAYER NEO

Davide Maiorca et al. [30], in 2015, proposed a model called Slayer NEO to detect the malevolent PDF archives by recovering the data from the content and inward structure from the investigated PDF record. Creators affirmed that this model had a productive parsing system revealed with it, which in turn will help with deciding the presence of non-JavaScript vindictive substance, too. With the organization of this order calculation, the authors express that their model has beaten different other static PDF malware detection systems [30]. The figure shows the architecture introduced for the Slayer NEO PDF malware location model. And to start with, the PDF document is gone through a component extraction module to draw out the presence of any implanted malwares [8]. For that reason, it utilizes parsers PeePDF and Origami [30]. After the include extraction, the various designs of the records are isolated, and the element vector is produced to group the given record as an amiable or a malicious one [8] (Figure 7.6).

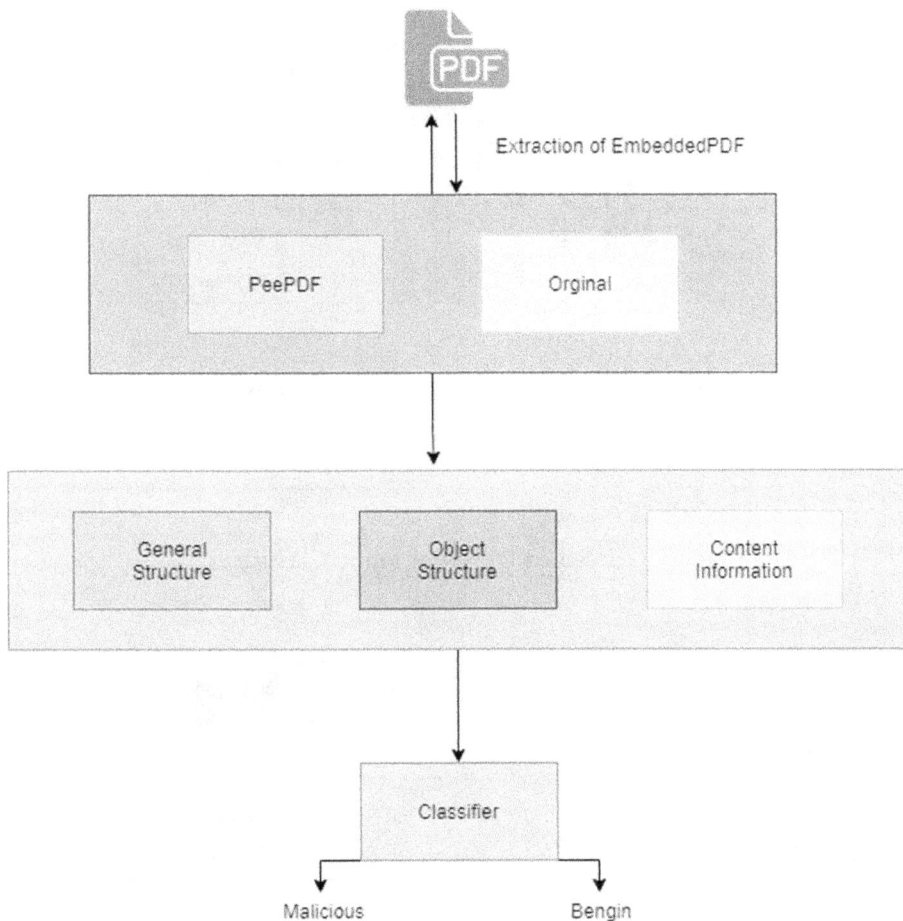

FIGURE 7.6 Architectural flow of Slayer NEO [8].

7.7.8 PDFᴛʀᴀᴛᴇ ᴠ2

Charles Smutz and Angelos Stavrou [11], in 2016, created a special and vigorous methodology that was fit for perceiving adversarial tests. For building this framework, the authors have tested with gigantic number PDF tests against various classifiers to recognize the particular scope of tests for which the "gathering classifier" was arranging lower efficacies [1]. At that point, dependent on the investigation of the attributes of those examples, the authors proposed an ensemble classifier, which worked on the premise of "common agreement analysis" strategy [1]. They likewise recommended that their methodology could be summed up to check different "angle descent" and "part density estimation assaults" (Figure 7.7 and Table 7.1).

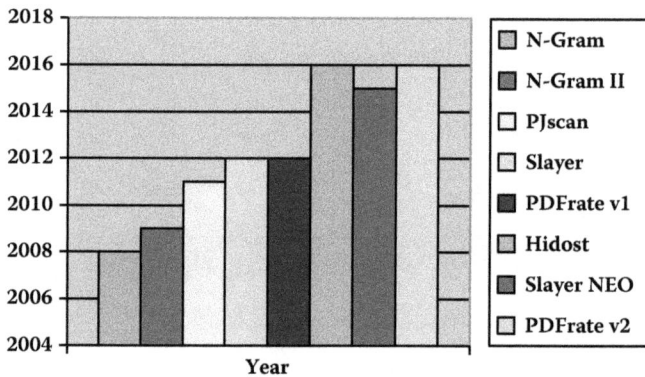

FIGURE 7.7   Year of implementation of different static classifiers.

TABLE 7.1

Type of Classifiers Used in Different Static Classifiers

Classifier	Author(s)
Markov	M. Zubair Shafiq et al. [32]
Decision Trees	S. Momina Tabish et al. [34]
SVM	Pavel Laskov and Nedim Šrndic [31]
Random Forests	Davide Maiorca et al. [17]
Random Forests	Charles Smutz and Angelos Stavrou [16]
Random Forests	Nedim Šrndic and Pavel Laskov [9]
Adaboost	Davide Maiorca et al. [8]
Classifier Ensemble	Charles Smutz and Angelos Stavrou [1]

7.8 COMPARISON BETWEEN THE DIFFERENT STATIC CLASSIFIERS

The above figures and graphs show us the different static classifiers with their year of implemenation and the classifier being used inside it.

7.9 DYNAMIC CLASSIFICATION OF PDF FILES

Dynamic identifiers are additionally, for the most part, centered around the JavaScript code of the record as their presence contributed to the raised doubt [4]. Be that as it may, this class of identifiers doesn't search for the examples or the frequencies of the catchphrases and different components [9]. All things being equal, it will straightforwardly remove these JavaScript code and attempt to execute them in strong disengaged conditions and look at their activities [31]. The encounter that was being looked at in the static finders because of training of concealing strategies can be defeated by conveying these sorts of detectors [28]. The dynamic detectors dissimilar to static doesn't look at the record just by doing an assessment on their inside structures [28]. They look at the nature of the document through extraction of blaze and scripting codes and execute them to get the document's genuine nature [5]. Even though these sorts of detectors are proficient to decide the malignant code proficiently, they may fizzle in instances of unsupportive conditions [11]. A portion of the JavaScript code may need some other extra component up-hold for doing its activities [16]. For that, we need to imitate the execution inter-action of the document substance by proffering it with its required necessities [11]. Another flaw can be that the dynamic detectors may bomb when the malwares are embedded into records without the use of JavaScript code [11] (Figure 7.8).

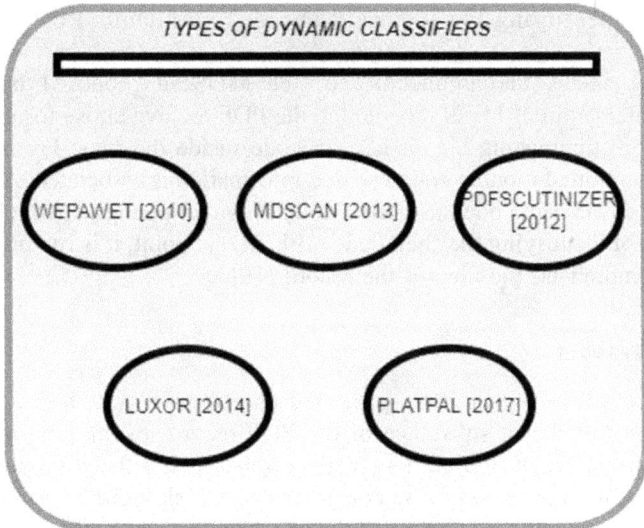

FIGURE 7.8 Dynamic PDF classifiers.

7.10   DYNAMIC CLASSIFICATION SYSTEMS

7.10.1   PlatPal

Meng Xu and Taesoo Kim [11], in 2017, prototyped a stage variety apparatus "PlatPal" which annexes to the "Adobe Reader" application to look at the inward components of the document under the parsing measure, and furthermore, it executes the document code in sandboxes to decide the impact of the code to the host's machine [11]. The primary guideline the authors used in building this framework was that a benign document would have comparative characteristics on execution regardless of the application used to execute it, while a malicious one could arrange different practices on handling by assorted applications [11]. This model is exceptionally adaptable and can distinguish the embedded malware with diminished "bogus alert" reactions [11].

In this, the PDF document is shipped off various virtual machines introduced with different operating systems to check its social properties at runtime. In the wake of running the document on these virtual machines, the outcomes obtained from these machines are looked at for the arrangement technique. With this cycle, the effect of the document can be recognized on different host stages.

7.10.2   MDScan

Zacharias Tzermias et al. [19], in 2011, proffered a framework that was able to break down the code genuinely and furthermore organized powerful preparation of the contained code inside the archive to perceive the malware dwelling in the PDF documents. On thinking about the framework in contrast to the informational indexes of PDF tests, the introduced model was seen to be equipped for identifying the intensely encoded pernicious PDF records [19] (Figure 7.9).

The figure shows the architecture of the MDScan proposed by Zacharias Tzermias et al. [19] in 2011. In this model, the PDF archive chose for investigation is parsed for distinguishing the JavaScript code inside the files. From both these modules, a committed module called shared information is associated. At that point, the identified JavaScript code has gone through JavaScript engine, which has an in-built element of identifying the shell code [19]. At that point, it is run on an imitated climate to comprehend the idea of the record [19].

7.10.3   WepaWet

Marco Cova et al. [33+], in 2010, proposed an effective system to do fastidious assessment on JavaScript substance of the PDF record. In this, the model fabricated and introduced by the authors will in general mix the discovery function with the imitating innovation to be equipped for perceiving the scripted code for playing out the analysis [33]. It helped the self-learning innovation to perceive the malware content covered up inside the JavaScript code through impersonating methods and contrasting the put-away anomaly designs [33]. The authors expressed that their model was additionally fit to add the recently recognized

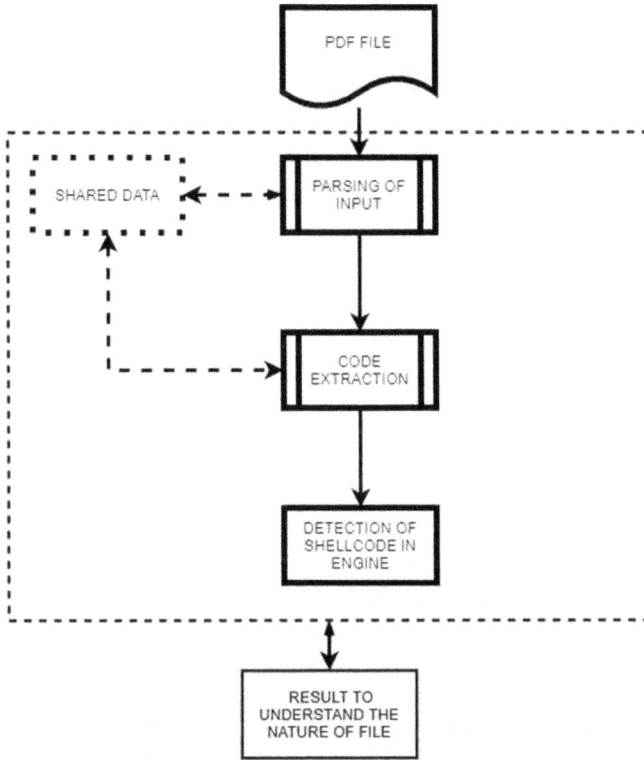

FIGURE 7.9 Work flow of MDScan.

malware marks itself, wiping out the need of unequivocal coding. Yen-Chen Lin et al. [6], in 2017, presented an ideal strategy to develop and implement the variety of adversarial attacks with the in-depth interpretation. A perfect set of actions to be executed by the systems are implicitly sequenced into the code, and these are tested against the custom attacks and inputted data for validation.

7.10.4 PDF SCRUTINIZER

Florian Schmitt et al. [15], in 2012, manufactured a dynamic PDF classifier named "PDF Scrutinizer," which was not just ready to group the PDF records proficiently, yet additionally proffered the thinking behind its dynamic status. The authors expressed that their model investigated all the presumed records that are installed with other.exe records independently from the ordinary documents that are not installed. This framework, in contrast to the different classifiers, ordered documents under any of the three classifications, which are malicious, benign, and dubious classes [15]. The PDF Scrutinizer used an interface called "PDF Box" for preparing the PDF records during the assessment [15].

The figure shows the functionalities and engineering of the PDF Scrutinizer model introduced by Florian Schmitt et al. [15] in 2012. The PDF Scrutinizer has

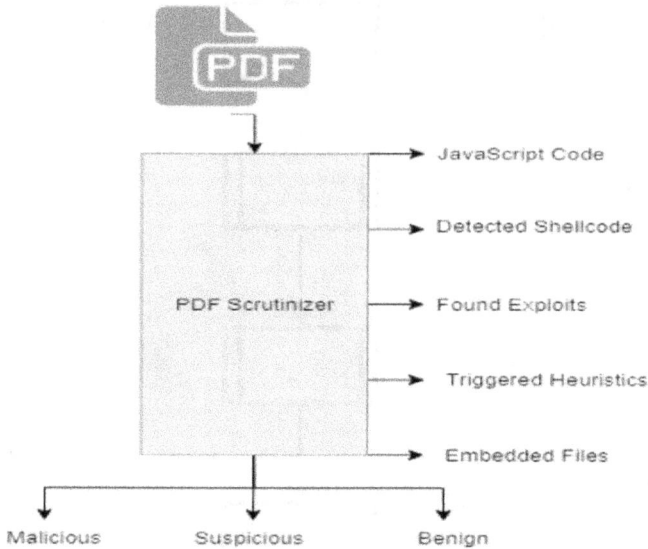

FIGURE 7.10 Working structure in PDF Scrutinizer.

various functionalities, as we can see from Figure 7.10. The PDF document, which we give to the framework, experiences different levels to check the prospects of malignant substance inside it [15]. The first is to discover the presence of JavaScript code inside the record under review [15]. At that point, we notice the shell code existing in it. Based on the finding, a factual investigation is directed to order the record as noxious, favorable, or dubious [15].

7.10.5 LuxOR

Igino Corona et al. [28], in 2014, manufactured a Dynamic Classifier called "LuxOR," which represents Lux on discriminant references." The model worked dependent on a "lightweight" technique to investigate the nature of the JavaScript code by thinking about the "programming interface references," which incorporate various capacities, factors, and different components pronounced in the record [28]. LuxOR chose the code present inside the JavaScript part of the records and removed its "programming interface" components to investigate them with the assistance of "AI" algorithms [28]. This technique was viewed as a proficient procedure to counteract against the "mimicry attacks" [28] (Figure 7.11).

The above figure shows the architecture of the dynamic-learning-based PDF malware classifier LuxOR proposed by Igino Corona et al. [28] in 2014. In this, rather than reviewing the entire PDF report, just the JavaScript code is investigated. When the JavaScript code is extricated from the document, that code is sent for separating the API references [28]. From that point onward, in view of API reference determination, the PDF is arranged on the following stage, which is a classification module [28] (Figure 7.12 and Table 7.2).

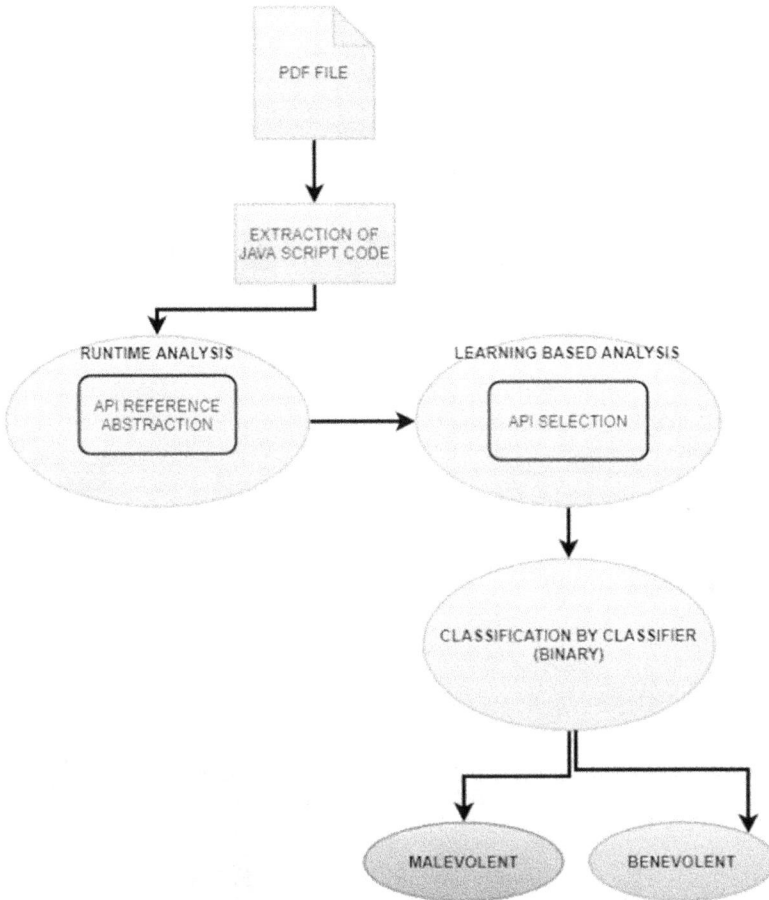

FIGURE 7.11 Flow in Lux0R.

7.11 COMPARISON BETWEEN DIFFERENT DYNAMIC CLASSIFIERS

The above table and figure depicts the comparison of the different dynamic classifiers we have taken into consideration for this paper. We can see the classifiers used for them and also these classifier systems are proved to be more efficient and prone to give more accurate results of the nature of the PDF files (Figures 7.13 and 7.14).

We also compared the true positive percentages of static classifiers with dynamic classifiers for figuring out the accuracy between them. We can see that the Slayer NEO is comparatively better and gives more accurate classification resuls to that of the PJScan.

7.12 NOVEL METHODOLOGIES

As the part of this work, we have gathered samples of malevolent and benign PDF samples from different sources and made a data set containing 1,400 samples of benign PDF and 1,500 samples of malevolent PDF files. The malevolent code inside the

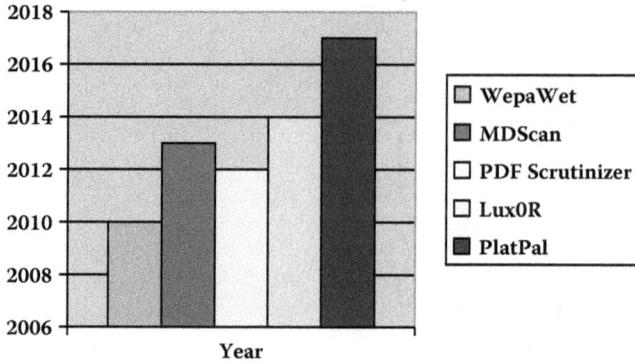

FIGURE 7.12   Year of implementation of different dynamic classifiers.

TABLE 7.2
Type of Classifier Used in Different Dynamic Classifier

Classifier	Author(s)
Bayesian	Marco Cova et al. [33]
–	Zacharias Tzermias et al. [19]
–	Florian Schmitt et al. [15]
Random Forests	Igino Corona et al. [28]
–	Meng Xu and Taesoo Kim [11]

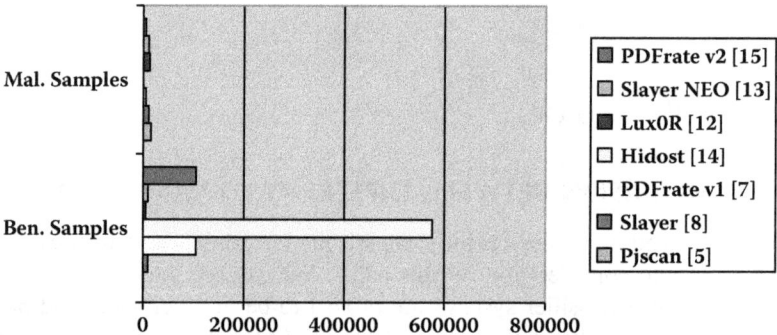

FIGURE 7.13   Samples taken.

malicious PDF files is so obfuscated that it can't be recognised as malicious by a user opening it. These PDF files are modified in their base structure to add malicious code so that the file can't be visible or recognised by a user having the document. We have developed a modified PDF classifier system on taking basis from the pre-existing "PlatPal" system. The system was coded in Python 2.7, so for the same, we have used a Python 2.7 interpreter. The other packages used in the system are numpy, pandas,

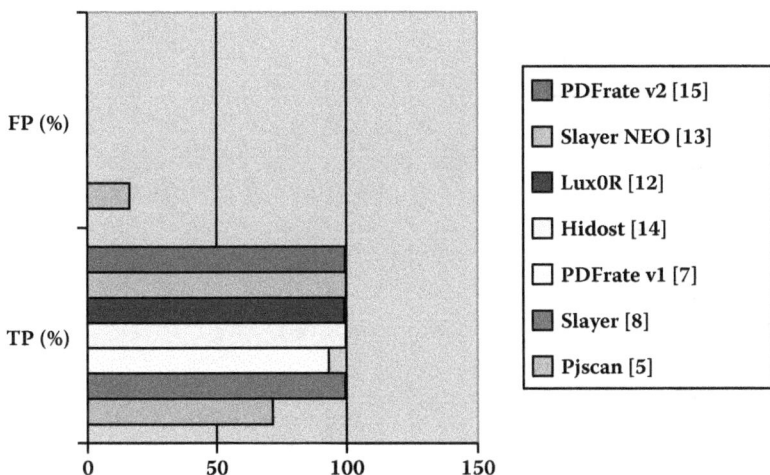

FIGURE 7.14   True positive percentage vs false positive percentage of the classifiers.

matplotlib, sklearn, scipy, xgboost, and PyWavelets. We used the Anaconda environment to design and execute the system. The system was trained with 80% of the total samples (benign and malicious) combined together and was tested for its accuracy as well as precision with the remaining 20% of the samples. Learning-based frameworks have been demonstrated to be defenseless against antagonistic information control assaults. These assaults have been concentrated under presumptions that the foe has certain information on either the objective model internals, its preparation data set, or else the classification scores it allocates to the samples under the inspection (Figure 7.15).

We need to calculate the accuracy and precision differences between the old system and the one we newly developed. Then we are going to make a code for implementing adversarial attacks on the fabricated system. In the testing phase, a particular sample is classified as a malevolent or a benign one based on its internal structure. But for making the classifier misclassify that particular sample set during the adversarial testing, we need to implement something that would be able to inject fabricated code into the file according to the nature of the file, so that during the adversarial attack, a benign PDF file should be classified as a malicious one, and, in the same way, a malicious PDF file should be classified as a benign one (Figure 7.16).

In our system, we have used different modules for training the system, extracting the features from the files while the system is getting trained, ETS (educational testing service) module, and the testing module, which can classify the given file with greater accuracy based on its extracted features.

In this thesis work, we will be fabricating an adversarial code to implement the adversarial attack on the dynamic PDF classifier system, which, in turn, would be successfully able to evade the system and make it misclassify the sample nature (Figure 7.17).

The above figure shows the main strategy we have implemented to classify the PDF file during the testing phase. It takes the references from the ETS generated during the testing procedure once the file is provided to the system. It extracts all the features

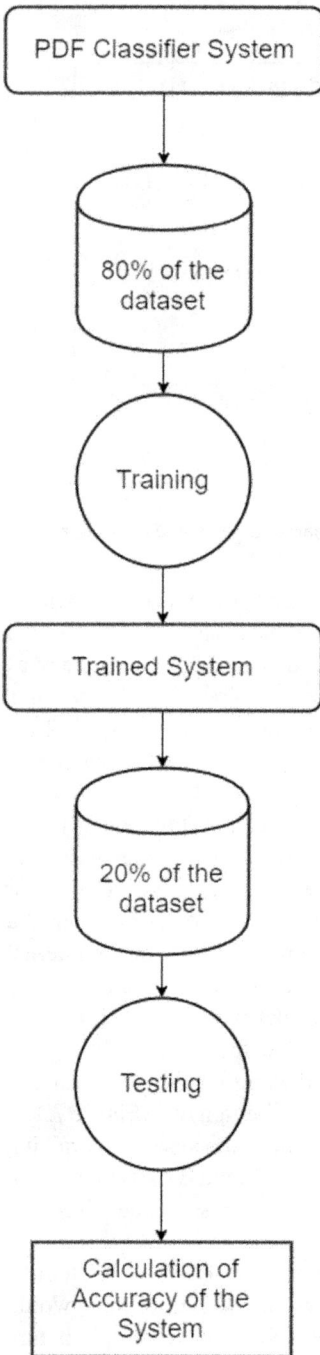

FIGURE 7.15 Workflow of the system.

Implementing Adversarial Attack in Testing phase

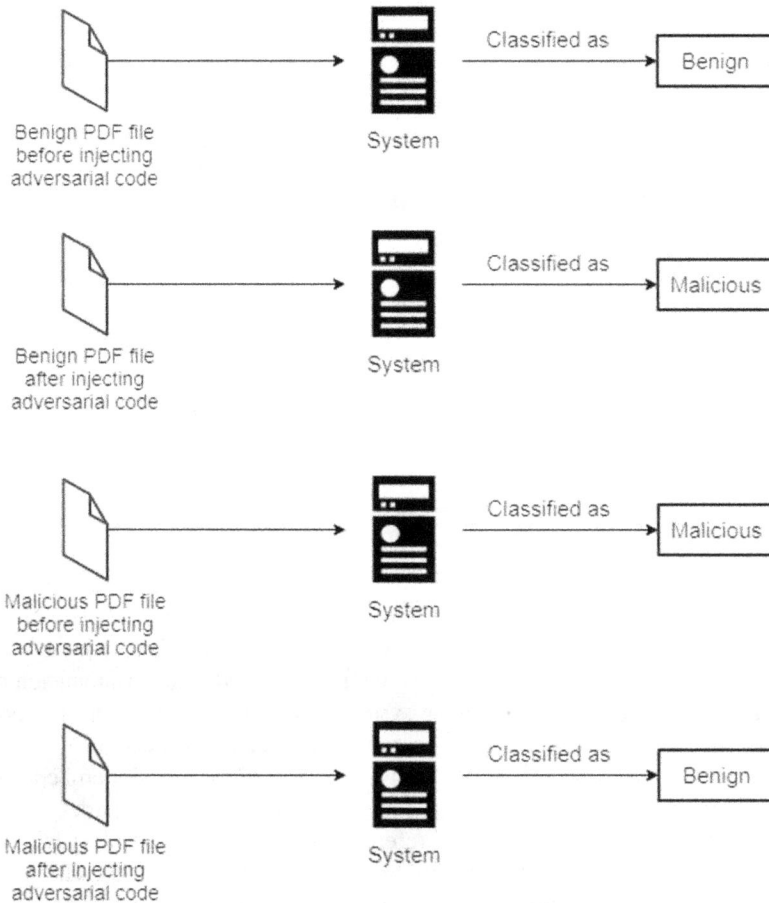

Classified as → Benign

Benign PDF file
before injecting
adversarial code

System

Classified as → Malicious

Benign PDF file
after injecting
adversarial code

System

Classified as → Malicious

Malicious PDF file
before injecting
adversarial code

System

Classified as → Benign

Malicious PDF file
after injecting
adversarial code

System

FIGURE 7.16 Implementing adversarial attacks during the testing phase.

from the given PDF file and tries to make the decision by taking references from the file. We need to further design a code that can make the file act as an adversarial sample to evade the classifier. This requires focusing on a uniform code that can make the required changes to the sample under inspection for making it evade the system.

7.13 CONCLUSIONS AND FUTURE DIRECTIONS

In this paper, we overviewed the various kinds of learning-based PDF malware classi-fiers. Fundamentally, the PDF classifiers are of two kinds, specifically, static and dy-namic classifiers. The static classifiers distinguish the indications of malwares by parsing through the entire archive. While the dynamic classifiers run the record in imitated conditions to review the practices of the document during the execution. The greater part of the dynamic classifiers fundamentally center around the JavaScript code in the record as the lion's share of the malignant examples have JavaScript content inserted into them.

```
codebook =np.loadtxt(open("center_6_250.csv","rb"),delimiter=",",skiprows=0)
while True:
    features=[]
    filePath = input("input the target file path(use exit to quit out):")
    if filePath.strip()=="exit":
        break
    fdata = open(filePath,'rb').read()
    ETS=EntrpyCala.generate(filePath)
    #ETS= EntrpyCala.generate(fdata)
    print('ets', ETS)
    features = Feature_extraction(codebook,ETS)
    if features==0:
        print("Error hanpend")
    features = tuple(features)
    Df = []
    Df.append(features)
    df2 =pd.DataFrame(Df,columns=columns)
    df2=pd.get_dummies(df2)
    X_test_std =scaler.transform(df2)
    flag = clf.predict(X_test_std)
    if flag[0]==1:
        print("malicious")
    elif flag[0]==0:
        print("clean")
```

FIGURE 7.17 Testing procedure.

From the outset, we have introduced the design of the PDF documents and how they are parsed utilizing a parsing programming. Following that, we talked about the designs of various static and dynamic classifiers. At that point, we led a near examination on the reviewed models. The features, grouping system, true positive (TP) rate, false positive (FP) rate, and F1 score were looked at among the changed classifiers.

We can also create more obfuscated adversarial PDF samples by embedding the malicious image file inside a PDF file object inside of the body part. The image object can be removed from the cross-reference table so that the object will not be seen by the classifier, even at the dynamic runtime inspection process. The classifier needs to be strengthened with more advanced techniques to read even the most obfuscated part for correctly classifying these kinds of variants. Hybrid classification systems would be recommended for these samples, but the time constraint should be taken care of side by side.

REFERENCES

[1] Lu, X., Zhuge, J., Wang, R., Cao, Y., & Chen, Y. (2013). De-obfuscation and detection of malicious PDF files with high accuracy. Proceedings of the Annual Hawaii International Conference on System Sciences, 4890–4899. 10.1109/HICSS.2013.166.
[2] Almrezeq, N. (2021). Cyber security attacks and challenges in Saudi Arabia during COVID-19. *Turkish Journal of Computer and Mathematics Education (TURCOMAT)*, 12(10), 2982–2991.
[3] Rani, P., Kavita, Verma, S., & Nguyena, N. G. (2020). Mitigation of black hole and gray hole attack using swarm inspired algorithm with artificial neural network. *IEEE Access*, 8. 10.1109/ACCESS.2020.3004692.
[4] Zhang, J. (2018). MLPdf: An effective machine learning-based approach for PDF malware detection. *arXiv*, 1–6. Retrieved from http://arxiv.org/abs/1808.06991.

[5] Kang, A. R., Jeong, Y. S., Kim, S. L., & Woo, J. (2019). Malicious PDF detection model against adversarial attack built from benign PDF containing javascript. *Applied Sciences (Switzerland)*, 9(22). 10.3390/app9224764.

[6] Lin, Y. C., Hong, Z. W., Liao, Y. H., Shih, M. L., Liu, M. Y., & Sun, M. (2017). Tactics of adversarial attack on deep reinforcement learning agents. Proceedings of the IJCAI International Joint Conference on Artificial Intelligence, 3756–3762. 10.24963/ijcai.2017/525.

[7] Biggio, B., & Roli, F. (2018). Wild patterns: Ten years after the rise of adversarial machine learning. *Pattern Recognition*, 84, 317–331. 10.1016/j.patcog.2018.07.023.

[8] Almadhoor, L. (2021). Social media and cybercrimes. *Turkish Journal of Computer and Mathematics Education (TURCOMAT)*, 12(10), 2972–2981.

[9] Srndic, N., & Laskov, P. (2013). Detection of malicious PDF files based on hierarchical document structure. Proceedings of the 20th Annual Network & Distributed Systems Symposium. Retrieved from http://scholar.google.com/scholar?hl=en&btnG=Search&q=intitle:Detection+of+Malicious+PDF+Files+Based+on+Hierarchical+Document+Structure#0.

[10] Schmitt, F., Gassen, J., & Gerhards-Padilla, E. (2012). PDF scrutinizer: Detecting JavaScript-based attacks in PDF documents. Proceedings of the 2012 10th Annual International Conference on Privacy, Security and Trust, PST 2012, 104–111. 10.1109/PST.2012.6297926.

[11] Batra, I., Verma, S., Kavita, Ghosh, U., Rodrigues, J. J. P. C., Nguyen, G. N., Sanwar Hosen, A. S. M., & Mariappan, V. (2020). Hybrid logical security framework for privacy preservation in the green Internet of Things. *MDPI-Sustainability*, 12(14), 5542. 10.3390/su12145542.

[12] Nissim, N., Cohen, A., Moskovitch, R., Shabtai, A., Edri, M., BarAd, O., & Elovici, Y. (2016). Keeping pace with the creation of new malicious PDF files using an active-learning based detection framework. *Security Informatics*, 5(1), 1–20. 10.1186/s13388-016-0026-3.

[13] Li, M., Liu, Y., Yu, M., Li, G., Wang, Y., & Liu, C. (2017). FEPDF: A robust feature extractor for malicious PDF detection. Proceedings of the 16th IEEE International Conference on Trust, Security, and Privacy in Computing and Communications, 11th IEEE International Conference on Big Data Science and Engineering and 14th IEEE International Conference on Embedded Software and Systems, Trustcom/BigDataSE/ICESS 2017, 218–224. 10.1109/Trustcom/BigDataSE/ICESS.2017.240.

[14] Vijayalakshmi, B., Ramar, K., Jhanjhi, N. Z., Verma, S., Kaliappan, M., Vijayalakshmi, K., Kavita, Ghosh, U., & Vimal, S. (2020). An attention based deep learning model for traffic flow prediction using spatio temporal features towards sustainable smart city. *IJCS,Wiley*. 10.1002/dac.4609.

[15] Nawaz, A. (2021). Feature engineering based on hybrid features for malware detection over Android framework. *Turkish Journal of Computer and Mathematics Education (TURCOMAT)*, 12(10), 2856–2864.

[16] Tzermias, Z., Sykiotakis, G., Polychronakis, M., & Markatos, E. P. (2011). Combining static and dynamic analysis for the detection of malicious documents. Proceedings of the 4th Workshop on European Workshop on System Security, EUROSEC'11. 10.1145/1972551.1972555.

[17] Laskov, P., & Šrndić, N. (2011). Static detection of malicious JavaScript-bearing PDF documents. Proceedings of the ACM International Conference Proceeding Series, 373–382. 10.1145/2076732.2076785.

[18] Smutz, C., & Stavrou, A. (2012). Malicious PDF detection using metadata and structural features. Proceedings of the ACM International Conference Proceeding Series, 239–248. 10.1145/2420950.2420987.

[19] Maiorca, D., Giacinto, G., & Corona, I. (2012). A pattern recognition system for malicious PDF files detection. Proceedings of the *International Workshop on Machine Learning and Data Mining in Pattern Recognition*, 510–524. Retrieved from https://link.springer.com/content/pdf/10.1007%2F978-3-642-31537-4_40.pdf.

[20] Humayun, M., Niazi, M., Jhanjhi, N. Z., Alshayeb, M., & Mahmood, S. (2020). Cyber security threats and vulnerabilities: A systematic mapping study. *Arabian Journal for Science and Engineering*, 45(4), 3171–3189.

[21] Maiorca, D., Biggio, B., & Giacinto, G. (2018). Towards adversarial malware detection: Lessons learned from PDF-based attacks. *arXiv*, *1*(1). Retrieved from http://arxiv.org/abs/1811.00830.

[22] Batra, I., Verma, S., Kavita, & Alazab, M. A lightweight IoT based security framework for inventory automation using wireless sensor network. *IJCS, Wiley*, 33(2), e4228. 10.1002/dac.4228.

[23] Chen, Y., Wang, S., She, D., & Jana, S. (2019). *On training robust PDF malware classifiers. arXiv*. Retrieved from http://arxiv.org/abs/1904.03542.

[24] Pitropakis, N., Panaousis, E., Giannetsos, T., Anastasiadis, E., & Loukas, G. (2019). A taxonomy and survey of attacks against machine learning. *Computer Science Review*, 34, 100199. 10.1016/j.cosrev.2019.100199.

[25] Song, W., Li, X., Afroz, S., Garg, D., Kuznetsov, D., & Yin, H. (2020). Automatic generation of adversarial examples for interpreting malware classifiers. *arXiv*. Retrieved from http://arxiv.org/abs/2003.03100.

[26] Goodfellow, I. J., Shlens, J., & Szegedy, C. (2015). Explaining and harnessing adversarial examples. Proceedings of the 3rd International Conference on Learning Representations, ICLR 2015, 1–11.

[27] Bauman, E., Lin, Z., & Hamlen, K. W. (2018). *Superset Disassembly: Statically Rewriting x86 Binaries Without Heuristics*. NDSS. 10.14722/ndss.2018.23300.

[28] Liu, Q., Li, P., Zhao, W., Cai, W., Yu, S., & Leung, V. C. M. (2018). A survey on security threats and defensive techniques of machine learning: A data driven view. *IEEE Access*, 6, 12103–12117. 10.1109/ACCESS.2018.2805680.

[29] Šrndić, N., & Laskov, P. (2016). Hidost: A static machine-learning-based detector of malicious files. *EURASIP Journal on Information Security*, 2016(1), 1–20. 10.1186/s13635-016-0045-0.

[30] Maiorca, D., Ariu, D., Corona, I., & Giacinto, G. (2015). A structural and content-based approach for a precise and robust detection of malicious PDF files. Proceedings of the 1st International Conference on Information Systems Security and Privacy, 27–36. 10.5220/0005264400270036.

[31] Cova, M., Kruegel, C., & Vigna, G. (2010). Detection and analysis of drive-by-download attacks and malicious JavaScript code. Proceedings of the 19th International Conference on World Wide Web, 281–290. 10.1145/1772690.1772720.

[32] Shafiq, M. Z., Khayam, S. A., & Farooq, M. (2008). Embedded malware detection using Markov n-Grams. *Lecture Notes in Computer Science (Including Subseries Lecture Notes in Artificial Intelligence and Lecture Notes in Bioinformatics)*, 5137 LNCS, 88–107. 10.1007/978-3-540-70542-0_5.

[33] Tabish, S. M., Shafiq, M. Z., & Farooq, M. (2009). Malware detection using statistical analysis of byte-level file content categories and subject descriptors. *Csi-Kdd*, 23–31. 10.1145/1599272.1599278.

[34] Hussain, S. J., Ahmed, U., Liaquat, H., Mir, S., Jhanjhi, N. Z., & Humayun, M. (2019). IMIAD: Intelligent malware identification for android platform. Proceedings of the 2019 International Conference on Computer and Information Sciences (ICCIS), 1–6. IEEE.

8 Key Authentication Schemes for Medical Cyber Physical System

*Zia ur Rehman, Saud Altaf, Saleem Iqbal,
Khalid Hussain, and Kashif Sattar*

CONTENTS

DOI: 10.1201/9780367808228-8

8.1 INTRODUCTION

The rapid development has been enduring in recent years in the various research areas, including hardware, communication technologies, software, etc. The embedded devices have played a pivotal part in various artificially intelligent applications. The extended research has transformed it to become a part of an emerging research direction called cyber physical systems (CPS). The CPS is one of an evolving trend in research that incorporates cyber, physical, and computing components together to unleash an innovative direction in the academic community. The subfields of CPS include smart grids, smart vehicles (unmanned ground vehicles (UGV)), unmanned arial vehicles (UAV), health monitoring, industrial control systems, and water/gas distribution networks. Many of these systems are deployed in critical environments and are playing a vital role in our daily lives.

Before expanding further, the difference between CPS and internet of things (IoT) needs to be elaborated. First, IoT connects sensing devices with the internet by supplementing features like high sensitivity, reliable transmission, and intelligent processing, while CPS utilizes information acquired from IoT with its ability to deeply incorporate the 3Cs (communication, computing and control). In other words, CPS extends the functionality of IoT, firstly by providing more robust control over physical entities with add-on safety, efficiency, and reliability, and secondly, it has far exceeded computational requirements for equipment compared with IoT.

Medical CPS is a special-purpose field devised from the health-monitoring subset of CPS. It requires sensing technologies for reliable data acquisition from remote locations, and for this purpose, it extensively depends on a specialized network called wireless sensors networks (WSN) for data acquisition, transmission, and control. The wearable or implantable devices are used for acquiring data from patients, which is then sent to medical practioner to diagnosis and further treat. The secure data transmission remains the paramount concern due to the openness of the communicational channel, which is wireless most of the time. The aim of the chapter is to address security objectives, numerous security challenges, and various categories of key authentication schemes as a promising solution to achieve those security objectives for the designers of medical CPS applications to enhance the quality of life. The architecture of medical CPS is shown in Figure 8.1.

The chapter is organized as follows: Section 8.2 addresses the security objectives, Section 8.3 highlights the potential security challenges, Section 8.4 describes the various types of key authentication schemes as a possible solution to security problems, and finally, the conclusion and future work is furnished at the end.

8.2 SECURITY OBJECTIVES FOR MEDICAL CPS

The following are the security objectives for medical CPS (Rehman et al., 2019):

8.2.1 MUTUAL AUTHENTICATION

Mutual authentical ensures that only authentic users can communicate among themselves and data communication is genuine. Therefore, it is vital requirement.

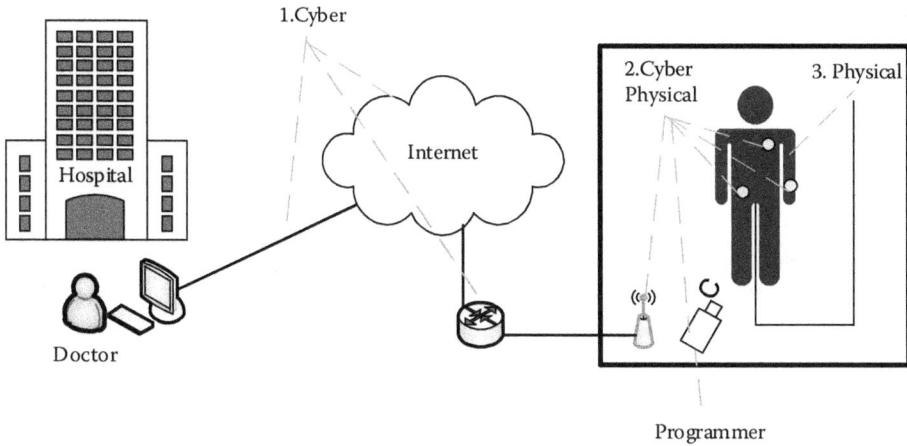

FIGURE 8.1 The architecture of medical CPS.

8.2.2 DATA FRESHNESS

This objective ensure that data is new and the system should be vigilant enough to identify older data. This feature helps to make sure that integrity and confidentiality of data is not disturbed.

8.2.3 FORWARD SECURITY

This security objective is to make sure that an entity cannot access or predict post-departure communication once it leaves the network. An adversary should not be able to predict any future secrets by guessing on the basis of some previous information.

8.2.4 DATA AUTHENTICITY

This objective makes sure that the valid data is communicated from a legitimate node and it is not from any intruder. Mostly the schemes use message authentication code (MAC) for this purpose.

8.2.5 DATA INTEGRITY

This objective ensures that the integrity of data is not disturbed due to any ilegi-timate modification during transmission. It makes sure that content of the message is not modified and is accurate.

8.2.6 DATA CONFIDENTIALITY

It refers to the fact data cannot be accessed by unauthorized persons, and it remains confidential. Medical devices have patients' data, which is extremely significant and should not be exposed to impostera.

8.2.7 UNFORGEABILITY

It refers to an important feature in which base station should not be compromised by an adversary in order to ensure smooth encumberance-free operation of any authentication scheme.

8.2.8 SCALABILITY

It refers to an important feature in which the authentication scheme should support network growth. In other words, the authentication process should not be disturbed or weakened with the introduction of new nodes.

8.3 SECURITY CHALLENGES FOR MEDICAL CPS

The following security challenges have been commonly found in literature.

8.3.1 RESOURCE CONSTRAINT (GUPTA ET AL., 2019)

It is a critical issue when collecting data from wearable or implantable devices to monitor a patient's condition. Therefore, the provision of security is challenging because traditional methods, such as assymetric crypto-solutions, are resource intensive and may not be suitable.

8.3.2 DENIAL OF SERVICE ATTACK (ALGULIYEV ET AL., 2018)

An adversary can overwhelm any device by sending extensive requests to make it unresponsive and thus make the device unavailable.

8.3.3 IMPERSONATION ATTACK (XU ET AL., 2019A,B)

An intruder tries to impersonate the nodes and acts like a genuine member of the network, but actually is not. This happens when an adversary has illegitimately gained access to the user's private credentials allowing them to behave like an actual user.

8.3.4 REPLAY ATTACK (KOMPARA ET AL., 2019)

An adversary somehow deceitfully captures the valid communication and replays it after some time to gain access to the network. It is a kind of attack that is difficult to detect because communication is normally valid and able to pass all the tests.

8.3.5 EAVESDROPPING ATTACK (SHEN ET AL., 2018)

An adversary tries to extract some useful information by listening to the data traffic to extract some secret that can be used to launch any successive attack.

8.3.6 Compromised Nodes and Clone Attack (Xu et al., 2019a,b)

In this attack, an intruder tries to compromise the most important node by gaining access to the particular node and commanding the node to behave as per its will. In a clone attack, it copies all the session keys to create a duplicate node that takes over the legitimate one.

8.3.7 Anonymous and Unlinkable Sessions (Kompara et al., 2019)

In this attack, an adversary tries to link two or more anonymous sessions to the same node to extract some valuable features, which can be used to unveil some secrets, like session keys, etc. Therefore, communication must be kept anonymous.

8.3.8 Desynchronization/Jamming Attack (Liu & Chung, 2017)

When the parties try to update their state in synchronization, after updating the state for the first party, the attacker compromises the communicating link so that the other party cannot update itself. Most of the authentication schemes are found to be vulnerable to this kind of attack.

8.4 TYPES OF KEY AUTHENTICATION SCHEMES FOR MEDICAL CPS

After identifying the security threats found in medical CPS, there is a dire need to develop a security mechanism that can adequately protect the communication among sensor nodes (N) and hub nodes (HN). The key authentication schemes are considered as a defense mechanism to secure the communication and are discussed in the remaining section. The researchers have categorized the authentication schemes in different ways; however, the most renowned types are included in the following section and are named as physiological based, cryptographic based, channel based, and hybrid authentication schemes (Rehman et al., 2019; Kompara & Hölbl, 2018).

8.4.1 Physiological-Based Key Authentication Schemes

These key authentication schemes use bodily features of a patient, i.e., heart, pulse rates, blood pressure, electrocardiogram (ECG), etc., to develop and agree on a single shared key, and this key is used to authenticate the communication among devices. Elyazidi et al. (2018) proposed a scheme that collects the physiological features like (ECG, pedometer, blood glucose, body movement, etc.) through body sensor nodes and transmits it to CU (smartphone). The sensor nodes are authenticated on CU if the true (1) correlation exists between the sensor node and CU is otherwise false (0) for the other case. If both sensor nodes and CU are on the same body, then the correlation is 1 else 0 for the opposite case. That's how sensor nodes are authenticated accordingly. The results of accuracy and precision are 84%–94.5% and 88%–91.5%, respectively.

Wu et al. (2018) proposed multitask EEG-based authentication schemes joined with eye blinking. The authors of this scheme have employed 15 imposters for each user to add more information to train classifiers to make the authentication process further stable. Furthermore, they have conducted open-set authentication tests for additional imposters to simulate the practical environment, which adds an extra edge to this scheme over the previous EEG-based authentication schemes. The results of this scheme have shown an increase in accuracy from 92.4% to 97.6%. Moreover, the robust security and reliability of this scheme is shown by a mean false accepted rate (FAR) of 3.9% and a false rejected rate of (FRR) 3.87%, respectively.

Sammoud et al. (2020) proposed an electrocardiogram (ECG) based symmetric key authentication scheme. Their scheme is based on three entities, namely, sensors attached to patients' body, the central node (PDA or cell phone), and the medical server. It also offers better key entropy due time variant property of ECG signal and ensures key retrieval. It is based on two important phases; firstly, the parent-child symmetric key establishment phase is responsible for sharing pre-installed keys with every node except the root node, and lastly, child-child biometric-based key generation phase that allows nodes having common predessor can establish a secure channel to communicate. The comparison results showed that the scheme is energy efficient and robust.

The authors of another study proposed a new scheme, namely CABA that is based on continuous authentication. It utilizes noninvasive biomedical methods for user authentication. The wearable medical sensors (WMSs) were deployed to collect physiological features like blood pressure (arterial systolic blood pressure, arterial average blood pressure, arterial diastolic blood pressure), body temperature, heart rate, oxygen saturation and respiratory rate. CABA engages nine biomedical streams excluding blood glucose (BG), ECG, and EEG to authenticate users continuously. It utilizes a two-phase authentication model, i.e., enrollment phase and authentication phase. The results disclose that it is a lightweight scheme with less cost as compared to peers (Mosenia et al., 2017).

Zouka securely integrated the healthcare system with WMS in his authentication scheme that enabled patients' health monitoring remotely. It utilized physiological features provided by WMS and stored in the healthcare system. The doctor is also intimated about patients' condition via SMS or email, if required. The proposed authentication scheme followed a three-stage model, namely, registration, patient login, and the authentication phase. The results showed it to be a lightweight, efficient, and effective authentication scheme (Zouka, 2017).

8.4.2 CRYPTOGRAPHIC-BASED AUTHENTICATION SCHEMES

The cryptographic-based authentication schemes are further divided into two parts, pre-deployed (symmetric) key authentication schemes and asymmetric authentication schemes. However, pre-deployed key authentication schemes provide the highest performance efficiency compared with asymmetric authentication schemes.

8.4.2.1 Pre-deployed Authentication Schemes

One of the recent studies, Kompara et al. (2019), has proposed an anonymous, authentication, and key agreement scheme, which is an enhancement of an earlier work proposed by Liu and Chung (2017). This study has only two cryptographic operations: hash function and XOR operation are used, which has turned it into a lightweight scheme. The scheme of Liu and Chung (2017) is found vulnerable for the untraceability of communicating nodes. The study of Kompara et al. (2019) has fixed the said problem by retaining the computational complexity of the original scheme and minimizing the communication cost, but on the price of increased cost of storage.

Moreover, on scrutinizing Kompara et al. (2019) further, the authors of another study (Rehman et al., 2020) have identified three vulnerabilities, namely, sensor-node impersonation, base station, and, intermediate node (IN) compromise attacks. Rehman et al. (2020) provided an enhanced scheme by not only providing a solution for these vulnerabilities but also making architectural-level vital changes in the original scheme. Thus reduced the overall communicational cost, which was remarkably lower compared with not only the original scheme but also with peer schemes as well. The network model of Rehman et al.'s scheme is shown in Figure 8.2.

Similarly, another improvement of Liu and Chung (2017) has been proposed by (Chen et al., 2018). They have highlighted that the Li et al.'s scheme is vulnerable

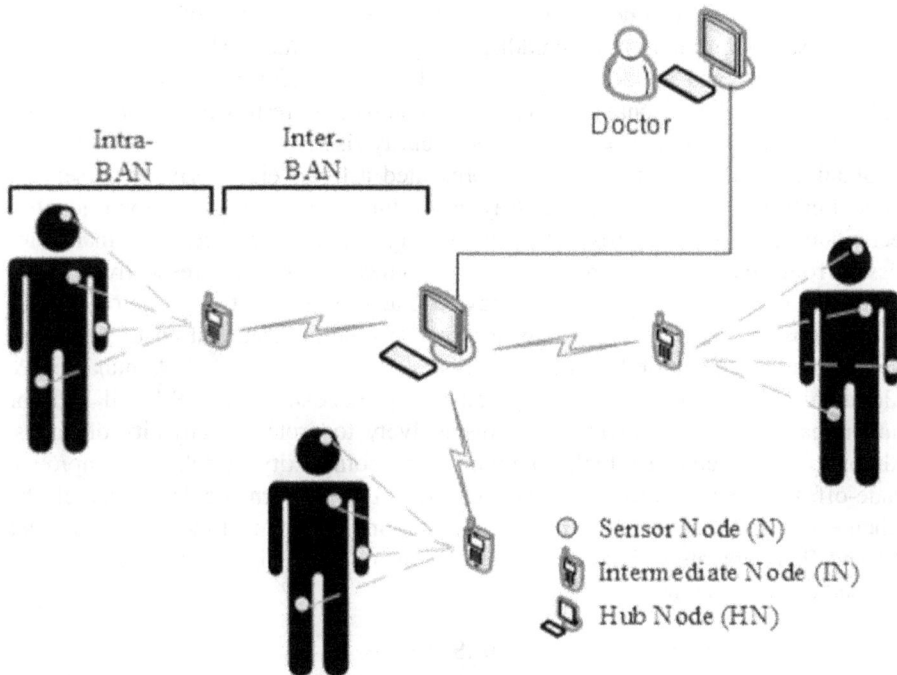

FIGURE 8.2   The network model of Rehman et al.'s scheme (2020).

to offline identity guessing, sensor-node impersonation, and hub node spoofing attacks. They have presented their own scheme by rectifying the highlighted flaws. Thus, they are providing more security and efficiency for the proposed scheme. They have modified the conventional way of communication between sensor node (SN), first node (FN), and hub node (HN) by limiting the role of FN as a relaying node because it is only forwarding data to HN without taking an active role in authenticating the nodes. This change has positively impacted the overall scheme by lowering the communication and storage cost over other peer schemes. Moreover, another enhancement of Li et al.'s (2017) scheme is also proposed by (Ostad-Sharif et al., 2019). They have pointed out that Li et al.'s scheme is vulnerable to wrong session key agreement and dsynchronization attacks. They have provided the solution to these problems, along with security against some other known attacks like impersonation, man-in-the-middle, modification, replay, hub-node stolen database, etc. The experimental results have shown better efficiency, security, and practicabillity.

In another study, (Xu et al., 2019b) has presented a lightweight, anonymous, authentication, and key agreement, which also relies on XOR operations and hash function. This study ensures forward secrecy without utilizing asymmetric encryption; even if an adversary some how compromises the master key. The experimental results of this study indicate low computational cost and also incur low security risk as compared with other peer lightweight schemes. Moreover, the lightweight scheme proposed by (Xu et al., 2019a) also uses XOR operations and hash function. The scheme provides security against a number of known attacks, e.g., eavesdropping, man-in-the-middle, replay, jamming/desynchronization, sensor-node anonymity and untraceability, forward/backward security, hub-node spoofing, and sensor-node impersonation attacks. The comparison results of this study depict that it has lower communication cost and security risk.

Shuai et al. (2020a) have recently presented a lightweight, privacy-perserving authentication scheme using one-way hash funchtion and pseudonym identity technique. The scheme provided resilience against forward secrecy, smart card loss, wrong password login and desynchronization attacks, respectively. The comparison results with peer work showed that it is efficient in terms of computational cost but slightly expensive in terms of communicational cost. Moreover, another enhancement of Kompara et al.'s work is proposed by (Almuhaideb & Alqudaihi, 2020). The authors proposed two protocols, namely P-I and P-II for authentication and re-authentication, respectively to protect anonymity of nodes. Although the scheme has high communication complexity, it achieved improved trade-off between performance and security. The performance analysis of the scheme showed that it has reduced computational cost and time if we consider P-I and P-II separately, but communication overhead is high on the hub node as compared with peer work.

8.4.2.2 Asymmetric Authentication Schemes

Shen et al. (2018) have proposedauthentication schemes for securing communication between personal digital assistance (PDA) and sensors, and the process used for key generation is kept lightweight. Another novel nonpairing, certificateless approach based one elliptic curve cryptography (ECC) is used for securing communication between PDA and the application provider (AP). The architectural design of this protocol is shown in Figure 8.1. The experimental results have shown the scheme is lightweight as compared to other RSA-based schemes; it is also efficient in a practical sense.

Xie et al. (2019b) presentated a certificateless authentication scheme, which is an improvement of an earlier work of (Ji et al., 2018) with conditional privacy-preserving (called CasCP). In this study, signature and authentication procedures are developed using elliptic curve cryptography (ECC), which eliminated complex bilinear pairing operation used previously. CasCP provides security against known attacks along with forgery and batch authentication attacks. The comparative results of this scheme have proved its advantage over the same schemes in computation and communication cost. Similarly, Xie et al. (2019a) proposed an improved certificateless aggregation signature scheme (iCLAS) based on a prior study of (Kumar et al., 2018). The iCLAS utilizes ECC, which makes use of an efficient message-signing algorithm, signature verification algorithm, and aggregation algorithm. The proposed scheme is also found resistant against known attacks and upon comparative analysis with other schemes; it is found better in terms of computational and communicational costs.

Truong et al. (2020) presented a chebyshev polynomial based authentication scheme in a multiserver environment. The scheme added a parallel session feature, which was not addressed in the peer work discussed in the article; it gave an extra edge over the peer work. However, storage authentication costs of the said scheme are comparatively higher than the related work. In another work proposed by Chaudhry et al. (2020) an improved and secure authentication scheme is presented. The scheme provided key agreement between cloud server through trusted authority and user. Although the said scheme is verified through a real or random (ROR) model, and informal security features analysis showed added advantage over peer work; but upon comparison of computation and communication costs, it resulted in a slight increase.

Moreover, Shuai et al. (2020b) proposed recently an authentication scheme using elliptic curve cryptography (ECC). The authors adoped identity-based certificateless authentication due to the proposed scheme becoming appropriate for multiserver architecture without participation of a third party. The comparative results showed that computational cost is slightly lower than the peer work, but it has high communicational cost, same as the storage cost. However, the scheme is better in terms of computational cost, and it is also privacy preserving.

In another research work proposed by (Shu et al., 2020), a certificateless aggregation scheme is used in conjunction with blockchain for secure storage of medical data. Although blockchain has limited capacity, it is used only for data-sharing purposes. The multi-trapdoor or chameleon function is used based on ECC to build a certificateless aggregation scheme. The experimental results showed that it is more computationally efficient than pairing-based schemes. The length of aggregate signature is constant, which means with increase of transactions, storage capacity remains same. This feature gave extra credit to it as compared to its peer work.

8.4.3 HYBRID AUTHENTICATION SCHEMES

The authors of a paper (Wazid et al., 2018) proposed a scheme that uses mobile password and biometric information to generate a key that is further utilized to authenticate the user. The proposed scheme is lightweight, and semantic security of the said scheme is proved by real-or-random (ROR) model, which showed that the scheme resists against the well-known attacks. It is also simulated through a NS2 simulation tool to measure the performance, and, upon comparison with other peer schemes, it showed low computational and communicational overhead.

Koya & P.P. (2018) proposed a hybrid scheme that improved an earlier work of (Liu & Chung, 2017) by combining it with electrocordiography (ECG) of the subject. The proposed scheme is designed in such way to improve the previous scheme by strengthening the vital security parameters. The biokey is generated from ECG signal by first calculating inter-pulse-interval (IPI); second by removing the most significant bits (MSB) and least significant bits (LSB), thirdly by gray coding, and lastly by applying concatenation operation. Hence, 128 bits biokey is generated through this process, which is used in authentication process. The performance and security analysis show that it provides more functionality and better security.

Challa et al. (2018) proposed a three-factor key agreement protocol and authentication to improve the limitations found in (Liu & Chung, 2017). The three-factors used in this study comprise user password, smart card, and user biometrics to improve the security, and this is the reason it is treated as a hybrid scheme. It is worth mentioning that three factors are combined with an ECC algorithm to make it low communication and low computation costs. This enhancement makes it feasible for wide applicability in the healthcare sector.

Chen (2020) presented biometric-based fuzzy authentication and key negotiation (BFAKN) scheme that provides overall authentication solution for all the entities involved in the communication process, including sensors, terminal, and the server. It deviced mutually agreed keys to ensure the validity of whole system, i.e., starting from local communication to terminal processing and then to server. The exprimental results showed that it provides high security, reliability, and effectiveness.

8.4.4 CHANNEL-BASED AUTHENTICATION SCHEMES

Zhang and Ma (2018) have proposed a key authentication scheme on the basis of similarity of received signal strength (RSS). The key is extracted from RSS values of sensor nodes and coordinator nodes. The inconsistency in the key is reduced using n-dimension quantification, and fuzzy extraction is used to transform the output of n-dimension quantifiers into a secret key. This study has high key entropy, which makes it more robust and difficult for an eavesdropper to guess the key. It also ensures low bit inconsistency rate and high key generation rate. The experimental results of scheme demonstrate it as a more feasible solution for resource constraint environment.

In paper (Zhang et al., 2018), the authors presented a variant of password authentication protocol without depending on preshared password assumption. The scheme extracts the password from fading channel and then usee it for devising a secret key for communication. Authors believed that their protocol would have better performance in terms of computation and communication, as compared to other password authenticated key exchange (PAKE) protocols.

Sciancalepore et al. (2019) presented EXCHANge protocol, which establishes crypto-less over-the-air key based on anonymity of sender/receiver. It is implemented in OpenWSN protocol and tested on OpenMote-CC2538 architecture. Another important aspect of this shceme is that it is the first real-world implementation of establishing a key protocol, which is based on the anonymity of the channel. It works on a physical layer and can set up the new key of 128 bits in an average of 10 seconds, although it requires only single hash function calculation. The experimental results have proved it secure against active and passive attacks and found it feasible for indoor scenarios and wearable applications.

Umar et al. (2020) recently presented authentication scheme based channel characteristics and an enhanced butterfly algorithm. The experimental results proved it an efficient scheme in terms of storage and computation costs. However, communication cost is slightly higher than one of the peer work. The scheme is also experimentally evaluated not only in different environments and scenarios but also in its performance, which is analyzed on different volunteers. The experimental results showed that the said scheme provided effective mutual authentication and resilience against various security attacks.

An authentication scheme name BANA was introduced by (Shi et al., 2013) that utilized variations of RSS to distinguish between legitimate and nonlegitimate nodes. The limitation of BANA was that the sensors need to maintain the line of sight (LOS), and it did not generate any secret key. This drawback was rectified in successive schemes, namely MASK-BAN (Shi et al., 2015) that extracted secret keys based on channel characteristics. The false positive rates are controlled using a multi-hop authentication mechanism. The experimental results showed it to be an efficient and effective authentication scheme.

In another research, authors utilized the ratio of RSS among '*wearable proxy devices (WPDs)*' and '*implantable medical devices (IMDs)*' to differentiate between a genuine user and an intruder. Furthermore, in order to improve IMD accessibility in emergency mode and to safeguard against forced authentication attack, they designed two '*authentication request filter (ARF)*' based protocols. The experimental results illustrated that the said scheme achieved a great authentication response rate of 99.2% for a genuine user and low authentication response rate for an intruder (Zhang et al., 2020).

Aman et al. (2020) proposed a lightweight protocol that used channel characteristics to excerpt wireless fingerprints and '*physical unclonable functions*' to achieve mutual authentication, anonymity, and data provenance. The scheme actually used a link-quality indicator (LQI) to distinguish among adversarial and nonadversarial wireless links. The experimental results showed that fingerprints accurately identify cyber-attacks, and it is energy efficient as compared to peer work.

Similarly, in another investigation, an authentication scheme was presented that utilized the user's behavioral fingerprints along with channel characteristics. The RSS is monitored for user actions like teeth brushing, water drinking, medicine taking, and breakfast eating in this experimental study. These behavioral data were collected and stored in a monitoring station that generated fingerprints for the user. This data is further utilized for the authentication process. The experimental results demonstrated that the scheme successfully recognized identity, even in worst case scenarios as well, and showed adequate resistance against intruders (Zhao et al., 2016).

8.5 CONCLUSION AND FUTURE RESEARCH DIRECTION

Medical CPS is a promising field that paves the way for additional research activities. It is a research arena with a lot of potential to improve quality of life in remote health-monitoring scenarios. However, the ease is accompanied by the cost, i.e., security of patient's data, which is of utmost importance. The new security challenges have been arising with the advent of new technology and interconnection of devices. Therefore, the severe concerns regarding security will be continuing. This chapter figured out the security objectives along with the challenges regarding security, which is found commonly. A summary of various types of key authentication schemes have been discussed briefly as a possible solution to these problem. By scrutinizing these schemes, it has been found that pre-deployed key authentication schemes are highly efficient, lightweight, and recommended choice for this field as compared with other authentication schemes. However, other authentication schemes are equally important and are potential candidates for deploying them in resource and energy constraint environments; still more research trends can be found in symmetric key crypto-solution and especially in pre-deployed key authentication schemes.

We perceive that the research regarding security of medical CPS will remain enduring in the future, as with the advancement in technology and security challenges. Channel-based schemes are found more robust and have extra edge in terms of key entropy, but they are computation intensive like asymmetric crypto-solutions. As discussed earlier, pre-deployed key authentication schemes are the potential candidate for future work due to lightweightness and low computation demands. Hybrid authentication schemes are another potential candidate for future work due to the robustness it offers.

REFERENCES

[1] Alguliyev, R., Imamverdiyev, Y., & Sukhostat, L. (2018). Cyber-physical systems and their security issues. *Computers in Industry*, *100*(April), 212–223. 10.1016/j.compind.2018.04.017.

[2] Aman, M. N., Basheer, M. H., & Sikdar, B. (2020). A lightweight protocol for secure data provenance in the internet of things using wireless fingerprints. *IEEE Systems Journal*, 1–11. 10.1109/jsyst.2020.3000269.

[3] Almuhaideb, Abdullah M., & Alqudaihi, Kawther S. (2020). A Lightweight and Secure Anonymity Preserving Protocol for WBAN. *IEEE Access*, 8, 178183–178194. 10.1109/access.2020.3025733.

[4] Challa, S., Das, A. K., Odelu, V., Kumar, N., Kumari, S., Khan, M. K., & Vasilakos, A. V. (2018). An efficient ECC-based provably secure three-factor user authentication and key agreement protocol for wireless healthcare sensor networks.

Computers and Electrical Engineering, 69, 534–554. 10.1016/j.compeleceng. 2017.08.003.

[5] Chaudhry, Shehzad Ashraf, Shon, Taeshik, Al-Turjman, Fadi, & Alsharif, Mohammed H. (2020). Correcting design flaws: An improved and cloud assisted key agreement scheme in cyber physical systems. *Computer Communications*, 153, 527–537. 10.1016/j.comcom.2020.02.025.

[6] Chen, C., Xiang, B., Wu, T., & Wang, K. (2018). An anonymous mutual authenticated key agreement scheme for wearable sensors in wireless body area networks. *MDPI*, 8(1074), 1–15. 10.3390/app8071074.

[7] Chen, Hanlin, Ding, Ding, Su, Shuchun, & Yin, Jingyi (2020). Biometrics-based cryptography scheme for E-Health systems. *Journal of Physics: Conference Series*, 1550, 022039. 10.1088/1742-6596/1550/2/022039.

[8] Elyazidi, Saâd, Escamilla-Ambrosio, Ponciano Jorge, Gallegos-Garcia, Gina, & Rodríguez-Mota, Abraham (2018). Accelerometer Based Body Area Network Sensor Authentication, *Lecture Notes of the Institute for Computer Sciences, Social Informatics and Telecommunications Engineering,Smart Technology* (pp. 151–16410.1007/978-3-319-73323-4_15.

[9] Gupta, A., Tripathi, M., Shaikh, T. J., & Sharma, A. (2019). A lightweight anonymous user authentication and key establishment scheme for wearable devices. *Computer Networks*, 149, 29–42. 10.1016/j.comnet.2018.11.021.

[10] Ji, S., Gui, Z., Zhou, T., Yan, H., & Shen, J. (2018). An efficient and certificateless conditional privacy-preserving authentication scheme for wireless body area networks big data services. *IEEE Access*, 6, 69603–69611. 10.1109/ACCESS.2018. 2880898.

[11] Kompara, M., & Hölbl, M. (2018). Survey on security in intra-body area network communication. *Ad Hoc Networks*, 70, 23–43. 10.1016/j.adhoc.2017.11.006.

[12] Kompara, M., Islam, S. H., & Hölbl, M. (2019). A robust and efficient mutual authentication and key agreement scheme with untraceability for WBANs. *Computer Networks*, 148, 196–213. 10.1016/j.comnet.2018.11.016.

[13] Kumar, P., Kumari, S., Sharma, V., & Kumar, A. (2018). Sustainable computing: Informatics and systems A certificateless aggregate signature scheme for healthcare wireless sensor network. *Sustainable Computing: Informatics and Systems*, 18, 80–89. 10.1016/j.suscom.2017.09.002.

[14] Koya, Aneesh M., & P. P., Deepthi (2018). An Anonymous hybrid mutual authentication and key agreement scheme for wireless body area network. Computer Networks, 140, 138–151. 10.1016/j.comnet.2018.05.006.

[15] Liu, C. H., & Chung, Y. F. (2017). Secure user authentication scheme for wireless healthcare sensor networks. *Computers and Electrical Engineering*, 59, 250–261. 10.1016/j.compeleceng.2016.01.002.

[16] Li, Xiong, Ibrahim, Maged Hamada, Kumari, Saru, Sangaiah, Arun Kumar, Gupta, Vidushi, & Choo, Kim-Kwang Raymond (2017). An Anonymous mutual authentication and key agreement scheme for wearable sensors in wireless body area networks. *Computer Networks*, 129, 429–443. 10.1016/j.comnet.2017.03.013.

[17] Mosenia, A., Sur-Kolay, S., Raghunathan, A., & Jha, N. K. (2017). CABA: Continuous authentication based on BioAura. *IEEE Transactions on Computers*, 66(5), 759–772. 10.1109/TC.2016.2622262.

[18] Rehman, Z. U., Altaf, S., & Iqbal, S. (2019). Survey of authentication schemes for health monitoring: A subset of cyber physical system. In 2019 16th International Bhurban Conference on Applied Sciences and Technology, IBCAST 2019, pp. 653–660. IEEE. 10.1109/IBCAST.2019.8667166.

[19] Shuai, Mengxia, Xiong, Ling, Wang, Changhui, & Yu, Nenghai (2020). Lightweight and privacy-preserving authentication scheme with the resilience of

desynchronisation attacks for WBANs. *IET Information Security*, 14, 380–390. 10.1049/iet-ifs.2019.0491.

[20] Rehman, Zia Ur, Altaf, Saud, & Iqbal, Saleem (2020). An Efficient Lightweight Key Agreement and Authentication Scheme for WBAN. *IEEE Access*, 8, 175385–175397. 10.1109/access.2020.3026630.

[21] Ostad-Sharif, Arezou, Nikooghadam, Morteza, & Abbasinezhad-Mood, Dariush (2019). Design of a lightweight and anonymous authenticated key agreement protocol for wireless body area networks. *International Journal of Communication Systems*, 32, e3974. 10.1002/dac.3974.

[22] Sammoud, Amal, Chalouf, Mohamed Aymen, Hamdi, Omessaad, Montavont, Nicolas, & Bouallegue, Ammar (2020). A new biometrics-based key establishment protocol in WBAN: energy efficiency and security robustness analysis. *Computers & Security*, 96, 101838. 10.1016/j.cose.2020.101838.

[23] Sciancalepore, S., Oligeri, G., Piro, G., Boggia, G., & Di, R. (2019). EXCHANge: Securing IoT via channel anonymity. *Computer Communications*, *134*(February 2018), 14–29. 10.1016/j.comcom.2018.11.003.

[24] Shen, J., Chang, S., Shen, J., Liu, Q., & Sun, X. (2018). A lightweight multi-layer authentication protocol for wireless body area networks. *Future Generation Computer Systems*, 78, 956–963. 10.1016/j.future.2016.11.033.

[25] Shi, L., Li, M., Yu, S., & Yuan, J. (2013). BANA: Body area network authentication exploiting channel characteristics. *IEEE Journal on Selected Areas in Communications*, *31*(9), 1803–1816. 10.1109/JSAC.2013.130913.

[26] Shi, L., Yuan, J., Yu, S., & Li, M. (2015). MASK-BAN: Movement-aided authenticated secret key extraction utilizing channel characteristics in body area networks. *IEEE Internet of Things Journal*, 2(1), 52–62. 10.1109/JIOT.2015. 2391113.

[27] Shu, H., Qi, P., Huang, Y., Chen, F., Xie, D., & Sun, L. (2020). An efficient certificateless aggregate signature scheme for blockchain-based medical cyber physical systems. *Sensors (Switzerland)*, 5(2020), 1521.

[28] Truong, T., Tran, M., & Duong, A. (2020). Chebyshev polynomial-based authentication scheme in multiserver environment polynomial-based authentication scheme. *Security and Communication Networks, 2020.*

[30] Wazid, M., Das, A. K., & Vasilakos, A. V. (2018). Authenticated key management protocol for cloud-assisted body area sensor networks. *Journal of Network and Computer Applications*, *123*(September), 112–126. 10.1016/j.jnca.2018. 09.008.

[31] Umar, Mubarak, Wu, Zhenqiang, & Liao, Xuening (2020). Mutual Authentication in Body Area Networks Using Signal Propagation Characteristics. *IEEE Access*, 8, 66411–6642210.1109/access.2020.2985261.

[32] Wu, Q., Zeng, Y., Zhang, C., Tong, L., & Yan, B. (2018). An EEG-based person authentication system with open-set capability combining eye blinking signals. *Sensors (Switzerland)*, 18(2), 1–18. 10.3390/s18020335.

[33] Xie, Y., Li, X., Zhang, S., & Li, Y. (2019a). ICLAS: An improved certificateless aggregate scheme for healthcare wireless sensor networks. *IEEE Access*, 7, 15170–15182. 10.1109/ACCESS.2019.2894895.

[34] Xie, Y., Zhang, S., Li, X., Li, Y., Chai, Y., & Zhang, M. (2019b). CasCP: Efficient and secure certificateless authentication scheme for wireless body area networks with conditional privacy-preserving. *Security and Communication Networks, 13.* 10.1155/2019/5860286.

[35] Xu, Z., Xu, C., Chen, H., & Yang, F. (2019a). A lightweight anonymous mutual authentication and key agreement scheme for WBAN. *Concurrency Computation*, *31*(14), 1–12. 10.1002/cpe.5295.

[36] Xu, Z., Xu, C., Liang, W., Xu, J., & Chen, H. (2019b). A lightweight mutual authentication and key agreement scheme for medical internet of things. *IEEE Access*, 7(c), 53922–53931. 10.1109/ACCESS.2019.2912870.

[37] Zhang, P., & Ma, J. (2018). Channel characteristic aware privacy protection mechanism in WBAN. *Sensors*, *18*(8), 2403. 10.3390/s18082403.

[38] Zhang, Y., Xiang, Y., Wu, W., & Alelaiwi, A. (2018). A variant of password authenticated key exchange protocol. *Future Generation Computer Systems*, *78*, 699–711. 10.1016/j.future.2017.02.016.

[39] Zhang, Z., Xu, X., Han, S., Liang, Y., & Liu, C. (2020). Wearable proxy device-assisted authentication request filtering for implantable medical devices. In IEEE Wireless Communications and Networking Conference, WCNC, 2020 May. 10.1109/WCNC45663.2020.9120856.

[40] Zhao, N., Ren, A., Hu, F., Zhang, Z., Rehman, M. U., Zhu, T., ... Alomainy, A. (2016). Double threshold authentication using body area radio channel characteristics. *IEEE Communications Letters*, *20*(10), 2099–2102. 10.1109/LCOMM.2016.2597831.

[41] Zouka, H. A. El. (2017). An authentication scheme for wireless healthcare monitoring sensor network. In 2017 14th International Conference on Smart Cities: Improving Quality of Life Using ICT & IoT (HONET-ICT), pp. 68–73. 10.1109/HONET.2017.8102205.

9 Ransomware Attack: Threats & Different Detection Technique

Rakhi Seth, Aakanksha Sharaff,
Jyotir Moy Chatterjee, and NZ Jhanjhi

CONTENTS

9.1 INTRODUCTION

Malware is malicious software that is created specifically for disruption or damage the system and gain the unauthorized access to the system. There are different types of malware like virus, worms, spyware, and many others; all these types can be understood through Figure 9.1.

Ransomware is software or malware that is considered to be kind of malicious. The concept of ransomware is that the attacker demands from victim a particular amount (it can be in different form, such as Bitcoin) to restore the access to the data upon payment. Each ransomware attack may have different characteristics or behavior; it depends on the country where it originates from; Different hacking groups associated within the country may ask for different threatening schemes, and it can be done through different ways like phishing spam. There are different types of ransomware like NotPetya. There is also a variation in the ransomware, like Leak ware or doxware, in which the attacker threatens to publicize the data of the victim, most probably the sensitive data that may harm the reputation.

The ransomware timeline started from the year 1989 PC Cyborg, which was developed by Joseph L. Popp; he hacked the conference of AIDS held at WHO by sending approximately more than 18000 files that are infected by using a floppy disk. This ransomware is based on a simple algorithm called symmetric cryptography and decrypts the file as well. After 1989, next new ransomware attack

DOI: 10.1201/9780367808228-9

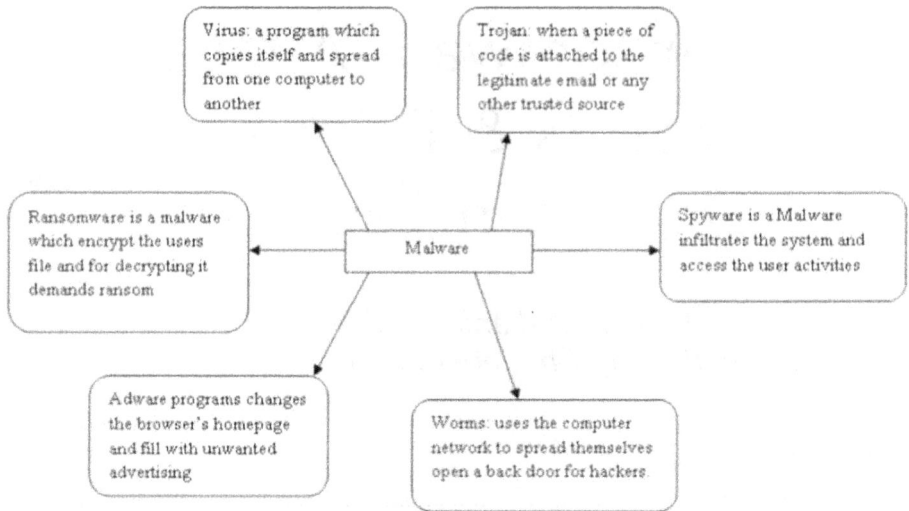

FIGURE 9.1 Types of malware.

evolution is done in 2005, which is known as Trojan. Gpcoder uses the spam email with the job application, and many of the earlier ransomware were were developed in Russia and aimed at the victim of neighboring countries. Next ransomware attack is generated in the year 2006 in which the concept is used that copied data files to password-protected archive files and stored them in the place of original files after deleting the original one. it is known as Trojan cryzipthere and is one more type that came in the year 2006; it is an improved version in that the ransomware attacker asks its victim to buy medication from specific pharmacies and submit the order ID when someone submit ID all data from system get copied. Locker ransomware, which is one of the types of RA attacks, started evolving in the year 2007 with a concept of displaying the pornographic image on the system; payment is asked for removing this malware from the machine, and SMS is used for unlocking code.

GP code is one of the malware with a 1024-RSA key, which is used to encrypt a file by the attacker side and exit the text file with the instruction at each point where it encrypts the file and has to pay a certain amount for decrypting the file [1]. This attack is generated in the year 2008. The burst of ransomware is another type of attack which came to light in the year 2011; it uses the concept of encrypting a file and demands money for decrypting it. One of the RA is Citadel, which attacks the system with the pay-per-install program; victims have to pay a minimum amount for that encrypted file, which the attacker already installed in the victim's system. It started in the year 2012. In the coming years, several other attacks came, as in the year 2013; an attack named Trojan ransom attacked the Windows security center and asked the victim to call on a particular phone number and get the control back over the system.

In the same year, one new attack came, which is known as the fake defenders. This one is specially designed for Android devices; it displays the threatening message and demands the payment to the victim and defender came into middle and solace to victim but twist is that defender is again one of the attacker and attack system more deeply. Trojan Sypeng is one of the models introduced in 2013; it was improved with the concept of locking the mobile phone by blaming the customer to be involved in child pornography. It comes from a legit source like the FBI and demands money from the victims.

Cryptolocker is one of the known RA attacks, and it uses the concept of encrypting and decrypting the files by using the private and public key. Firstly, it spread through the game and then it spreads through UPS or FedEx. Different types of Cryptolocker do the encryption on different files (nearly 67 files), and victims have to pay the amount within three days. In this method, an attacker uses the Bitcoin payment mode, and payment should be made through CashU or others. There is one more version that came from crypto locker, i.e., 2.0; it is also released in 2013 and implemented in a different language than language-implementing lockers.

Crypto wall is one of the improved versions and was introduced in the year 2014. In this version, harm will happen in a much more aggressive way than the earlier ransomware hacks; the files on the devices are harmed, as well as the secondary storage of the system being harmed, and it is connected to victims. The next iteration or version came in the year 2015 in which updated protocol is present so that detection of an attack is not made easily. Koler is one of the ransomware attacks that is made for encrypting the files present in Android devices; it also harms some of the buttons on the devices, and a bribe payment is asked through electronic mode. One more attack is presently known as Symplocker, in which the attacker locks the screen of the victim and demands the ransom. There are several others, like Lockri, Locker, and Petya. Petya works on making the hard disk inaccessible and demands money. Then Wannacry was introduced and attacked 150 countries, demanding thousands of dollars through 300,000 targets. Bad Rabbit is also introduced in the year 2017; this one targets the victim by refurbishing the victim system with the forged flash player to corrupt the files.

Gand Crab is malware that works on exploiting emails, and from time to time, aggravates the delivery by a phishing email; once it infects the victim's system properly, the attacker asks about the ransom, and it is a continuous process. Recently developed in the year 2019, a malware known as Dharma corrupted the files of the victim by encrypting them and asking for a ransom to decrypt the files. It uses the crypto virology for encrypting the victim's file. LockerGoga is also one of the attacks that are introduced in the year 2019 and computer or server files also get encrypted. SamSam is another malware that mainly focuses on an organization that works in the healthcare sector situated in various cities of the United States. This group got a benefit of $6 million as a bribe payment.

The above evolution can be understood briefly [2] because some of it might not be included in the above discussion. From Table 9.1, you get the idea of each attack.

Up until now, we understood the different ransomware attacks that may happen or that emerged over the period [3]. When we say how ransomware attacks work, then there is a whole cycle that came into action, and it will be done in many stages. This process can be explained through the following steps and Figure 9.2.

TABLE 9.1

Attack Evolved Over the Years

Attack	Year in which Attack Launched	Concept Behind the attack
PC Cybrog	1989	Floppy disk is used
GPCoder	2005–2008	Threads spread through emails.
WinLock	2010	Block the system and display the ransomware messages on PC
Reveton	2012	Purportedly warns from LEAs
DirtyDecrypt	2013	Encrypt some files
CryptLocker	September 2013	A Public key is taken from Command & Control
CryptoWall	November 2013	Only TOR browser is used for demanding a payment
Android Defender	2013	Android-based locker ransomware was introduced.
Torrent Locker	August 2014	Identical from SSH connection
CTB-Locker	December 2014	Uses bitcoin, elliptic curve cryptography, and TOR
CryptoWall 3.0	January 2015	Mainly uses the TOR for payment
TeslaCrypt	February 2015	It uses the PayPal cash cards for ransom
ClyptoWall 4.0	January 2015	Encrypt the file names
LinuxEncoder	November 2015	Works and encrypt the Linux home and any other directories.
DMA-Locker	January 2016	For decryption built-in features are used.
Locky Ransomware	February 2016	Malicious Marcos gets installed for a word document.
CTB-Locker (Websites)	February 2016	WordPress is used to target.
Carber	March 2016	It is a Malware factory
Petya	April 2016	Encrypt MFT and overrides MFT with its loader
CryptXXX	May 2016	Tracks the activities and avoids the sandboxed environment.
RAA	June 2016	It completely works on JavaScript
Satana	June 2016	It combines the features of MISCHA and PETYA
Stampado	July 2016	It mainly works on Dark Web
BadRabbit	2017	By using a fake flash player constant updation is done in the victim's system
Dharma	2019	Corrupting the files
LockerGoga	2019	Encrypt the files in server-side
SamSam	2019	Targets the health organization.

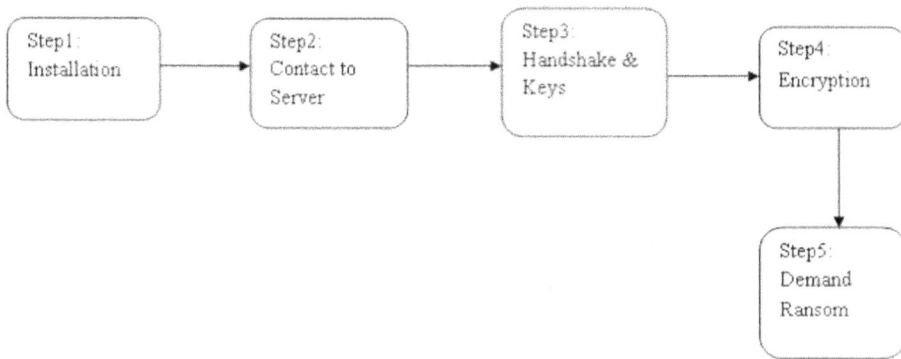

FIGURE 9.2   Stages of ransomware attack.

Step 1. **Installation:** It is one of the initial steps as well as the most important one; installation of an infected Trojan means that a file that contains the ransomware has been downloaded by the victim through an infected file; until then, e-stalking is done by the attacker. Once it is installed in the OS implicitly, without the user knowing that allocation of spyware is done. In windows OS, the Trojan sets the keys in the registry; this strategy attacker may attack each PC when it reboots. For mobile devices, phishing is done through unprotected apps. Once the installation is done, then that time ransomware marks and takes control over the victim's PC. Once it approves that the system is worth infecting, at that time, the spyware covers itself as one of the windows processes such as svchost.exe.

Step 2. **Contacting to Server/Headquarters:** In this stage, the victim's system, as well as the operating system, are working fine. The attacker contacts the source before beginning the attack. We can understand it in this way: the attacker is waiting for orders, and then the attacker will start his actions. Once the deleterious code is positioned and successfully installed, then it reaches out for orders because of various requests of a specific type. There are various communication channels for different variants and categories.

Step 3. **Handshake & Keys:** Handshake is a process that is different for each ransomware fraternity or owner. The CryLocker uses the PNG file on the real web page. Once the agreement is done between the client and server, the start of the next step is there, i.e., generation as well as an exchange of a key. Key exchange is done for enciphering in this step; it is done according to the complexity of the ransomware. The attacker may use the simple symmetric-key cipher. The key exchange executes, the private key is kept at the offender server, and the public key is installed in the prey's machine with the harmful code.

Step 4. **Encryption:** In this stage, encipherment is supposed to damage the files on the victim's machine, and, in this stage, only pernicious code becomes active. This code is associated with any file like Trojan,

MSOffice, GIF file, JPG, or another. Some of the crypto lockers en-
cipher files, as well as the file names, and then they start enciphering all
the files present in the system.

Step 5. **Demand Ransom:** In this stage, the attacker asks for ransom, or we can
call it extortion, for deciphering the files present in the victim's system.
In this stage, a unique HTML file is created and placed on the desktop;
in the case of crypto locker, the message arrives saying there is only one
way to restore the file, the private key. The victim gets the private key
when he/she will pay the above-mentioned amount. The amount can be
requested in various ways, like Bitcoin or asking a victim to buy the
medicine or health policy from a given source the attacker specifies.

As we saw the evolution of ransomware attacks over time now, we understand the
categorization based on different features associated with each type [4] of author
categorized, it into two types

1. Locker Ransomware
2. Crypto Ransomware

Crypto ransomware is an attack in which malware encrypts the victim's device
and demands payment for decrypting the files. The concept of how crypto ransomware work can be understood through [5]. It tells us about how the key is
generated and how the encryption will take place; then the public key is sent, and
the encrypted files are generated, as shown in Figure 9.3.

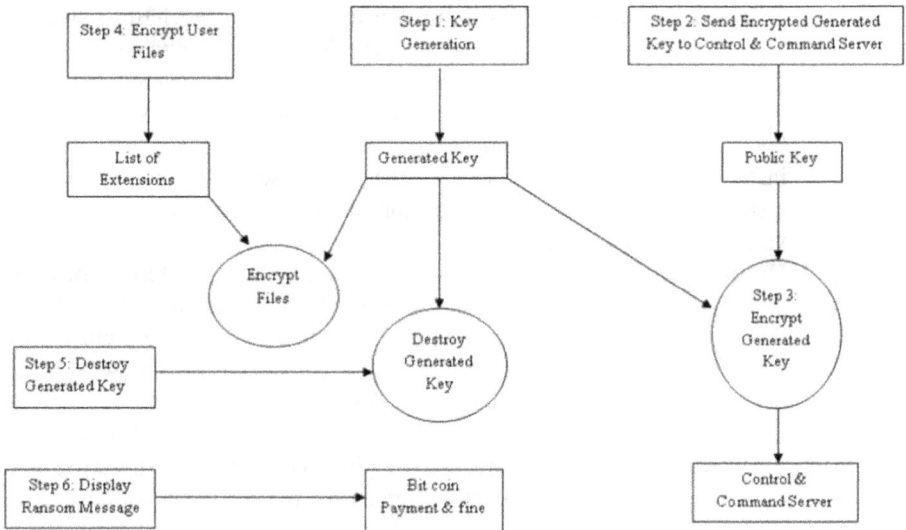

FIGURE 9.3   Components of crypto-ransomware.

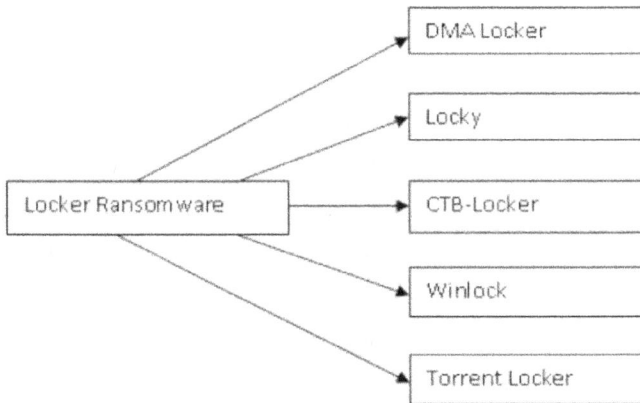

FIGURE 9.4   Different locker ransomware.

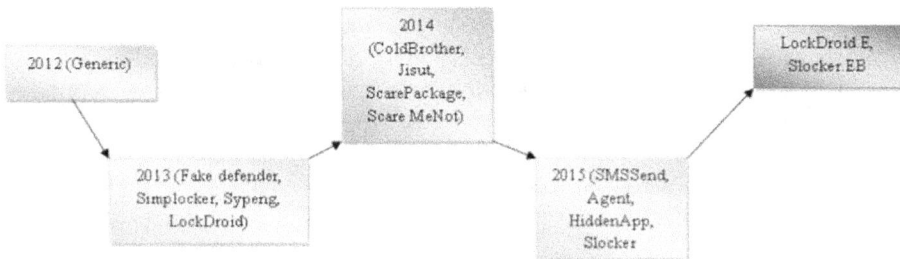

FIGURE 9.5   Time-based android ransomware.

Locker Ransomware:

In locker ransomware, the file is not encrypted; the attacker locks the victim's device and prevents from doing any tasks, and then the attacker demands the ransom. The user gets the pop-up window that shows the message of the virus in the system. The user may not attempt to shut down the system; otherwise, all the data will be lost. If a user tries to cancel the pop-up, it immediately returns, so there is no way out for the victim. By shifting the storage device, one can recover the data. There are various types of locker ransomware, as shown in Figure 9.4 above.

When we say locker ransomware, then it affects mostly Android devices. It first appears in 2012, and from 2012, it starts evolving. It can be understood through Figure 9.5.

Fakedefender is a fake antivirus that emphasizes purchasing. It provides the solution to discard malware from the victim's device, and, in reality, the malware is not present in the system. SimpLocker uses AES encryption to encrypt the files. Pletor encrypts the files on the Android device's memory card. Ransomware analysis occurs on Windows platform. All variants are analyzed using Cuckoo Sandbox and Anubis. After the analysis, reports and traffic files were observed thoroughly and main observations were documented. These changes can be observed in file system activities, registry activities, network communications, and locking mechanisms.

When we say an attack has happened, that means we detected the attack. There are various strategies through which victims can detect an attack [6,7]; as the author notes, there are various detection techniques that arose over the period of the evolution of attacks. The author [6] stated that anti-ransomware tool can prevent the damage that happened from the ransomware attack and also provide the recovering strategies for lost data. There are three major goals stated by anti-ransomware software.

1. To inhibit destruction to the system in question.
2. To detect formerly unseen ransomware.
3. To be hidden from the victim as much as possible.

Ransomware detection techniques are categorized into various types:

1. **Behavior-based:** This type detects the ransomware based on real-time changes that happen in the system, mainly in the user's file and data. It offers a specific file system aspect in which mainly crypto-ransomware works. For example, UNVEIL R-Locker is the detection technique that works on this concept.
2. **I/O Request Packet Monitoring:** RWGUARD is a type of packet monitoring. RWGUARD detects the crypto ransomware with three monitoring techniques:
   i.   decoy monitoring,
   ii. file change monitoring
   iii. process monitoring

   Various other examples are available, and they work on a similar concept. ShieldFS is an example, which works on both clean and infected machines with an I/O request packet.
3. **Network Traffic Monitoring:** As the name suggests, this technique analyzes the network traffic by monitoring the shared storage devices in a network to identify the ransomware. Here, a device detection algorithm is used to find the minimum no. of files deleted. The fundamental concern is the deletion and overwriting of files during the ransomware attack.
4. **At Storage Level:** It works on the flash-based storage device, and it can be done through the buffer management policy in which ransomware activity can find out through the read-and-write traces on the same location. Ex: ID3, SSD.
5. **API Call Monitoring:** GURLS is an example of API call monitoring; it uses the method that works on this feature, which is the frequency of API calls and strings.
6. **In Android devices:** It uses the hybrid approach, where the static and dynamic analysis for ransomware attacks a device. The static analysis focuses

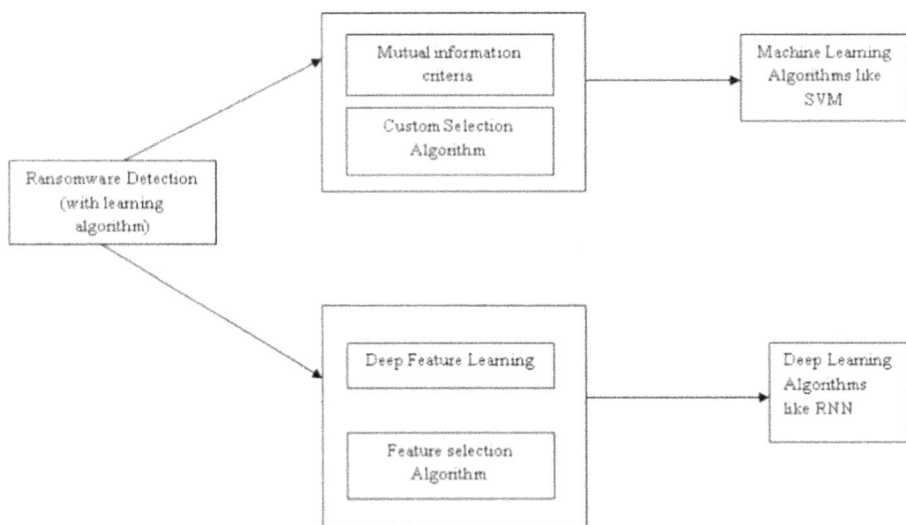

FIGURE 9.6 Ransomware detection with a learning algorithm.

on the opcode, and the dynamic analysis focuses on the memory; it is done by extracting the data from execution logs.

The author [7] uses the ransomware detection technique with a learning algorithm. This can be understood from Figure 9.6. When the learning algorithm is used, then the first criteria is the selection of features by using different algorithms and creating the optimal feature set. Then, training and testing is done by using this optimal mode; learning starts, and it can be done in different ways, such as supervised and unsupervised learning. The supervised learning algorithm uses the ransomware data samples as a training set and identifies the pattern of ransomware with a normal one to distinguish the two from each other. The unsupervised one fed the data set, which is not labeled, and attempted to find the pattern and build the model.

9.2   RELATED STUDY

This section gives the details of the technical methods that are used for preventing, and detecting the attack. The author said that [8] it uses the robust encrypting algorithm and key structure, but this is not the only quality of this ransomware attack. It also suggests the integration of hacking techniques that dropped by the shadow brokers. In other words, WannaCry is a combination of various modules that run together and attack the system. The author gives the thorough malware analysis through which the identification of malicious binary then collects the pattern in that malicious data; then, the data is preserved and the compromised situation is understood through indicators. It also reports the findings for the future strategies against the attacker. The author [9] tries to develop a generalized detection approach using predefined rules. The author extracts the traffic features by using the principal analysis, and all malicious content can easily be

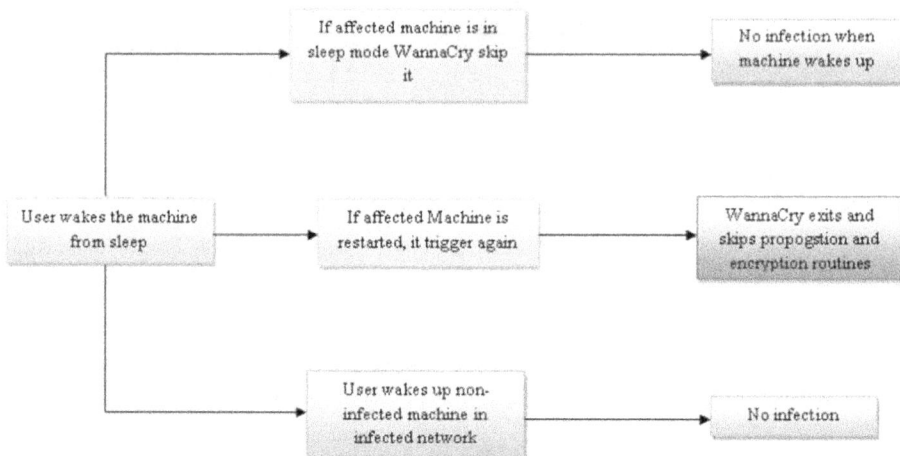

FIGURE 9.7   WannaCry protection strategy.

differentiated from the normal traffic in the middle of the attack. In this paper, the author went through eight stages by using the CWDR (comprehensive wannacry detection rules) rule set to detect WannaCry attacks. In Figure 9.7, we understood the protection strategies through which one can protect the system from the WannaCry attack.

Author [10] introduces the concept of a cyber kill chain. It is a path an adversary must take if an attacker wants to execute the attack. Every link in the kill plays a vital role, and if the link breaks, then the attack never happens. By removing any of these conditions, one can weaken the ransomware. Those conditions are specified by the author in the paper [10]; they are as follows

Condition 1: Entry and execution is the first condition; it means the attacker tries to enter into the host system to start infecting the system.

Condition 2: This condition tries to generate the unique random integer, and it is used to generate the encrypted key.

Condition 3: Successful ransom extraction is the third condition, which uses the decryption key.

Condition 4: For starting actual encryption, it requires the modification privileges.

Condition 5: The ransomware must be able to deny access (by the victim) of critical files, which requires privileges over the victim.

Condition 6: The last link is a functional payment route in which a path is there between the victim and the ransomware operator.

So the author proposes the NIST cybersecurity framework with five core functions. Wannacry is not the only type of ransomware attack; there is another one called the Petya ransomware attack [11]. The family of Petya attacks contains various other attacks, like PetrWrap and NotPetya. The concept of this type of attack is modifying the Windows system master boot record, which may crash the system. Petya uses the encryption key, i.e., AES-128 with RSA; the files can be encrypted without the system reload, and Petya also uses remote access to Windows management instruction using commands.

Until now, we understood about the ransomware attack type detection strategy, but ransomware attack origin starts when the concept of cybersecurity and malware detection came. These are the areas through which not only ransomware, but also the other types of attack, is detected. The author of [12] described the critical infrastructure that is used in cybersecurity. The important functions in the framework provided by NIST are as follows:

1. **Identify:** Identification of various cyber assets, as well as the risk analysis of exposure of this asset; moreover, it involves strong crypto-graphic authentication for both humans and equipment.
2. **Protect:** This function is defined as the protection mechanism used for access control, defense, zone division, authenticated communication, etc.
3. **Detect:** Detection is a function that calls for malware detection, intrusion detection, penetration testing, and vulnerability assessment.
4. **Respond:** This starts functioning when an attack happens or is activated and when an attack is detected. Then the first task is to control the current, as well as future, damage from the attack. This can be done by localization and containment.
5. **Recover:** These functions call out when someone wants to recover the data or files after the attack happens, as well as the planning of a better detection strategy through which the response function improves.

In this paper, [13] the author said that the malware detection system depends on the malware discriminated extracted features, which are evident through the analysis technique. Various analysis techniques work on static and dynamic tools. After extracting features, the system has to train through malware classifiers.

1. **Static Malware Analysis:** It works on the concept of feature extraction, without taking malware samples. Static features are extracted through different analysis functions, like hash value, opcodes, strings, and N-grams.

The author [5] uses the crypt ransomware STATIC, which takes the binary input and then generates the libraries. These files are DLL. Various static tools are there:

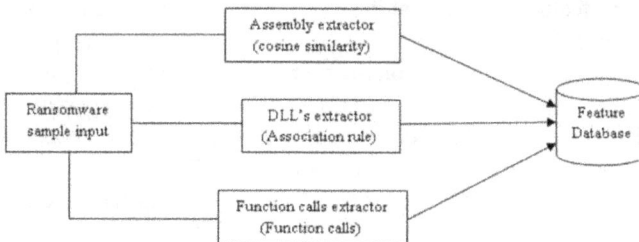

FIGURE 9.8 Static analysis of ransomware.

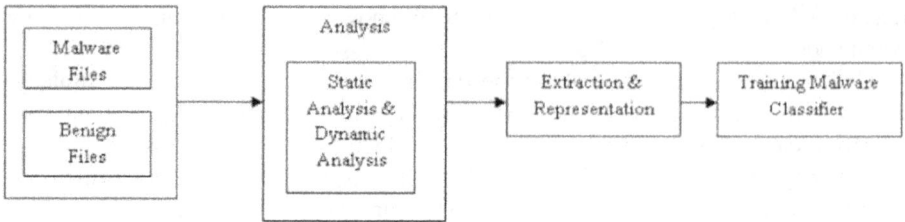

FIGURE 9.9   Training phase in malware detection.

PeView, PEid, CFF Explorer, Psfile, Accesschk. Through Figure 9.8, we see the working of static malware analysis.

2. **Dynamic Malware Analysis:** This works on the runtime, which means that malware files are executed, and in runtime, one can capture the activities of malware. It is based on the interaction of malware and computer system. A virtual environment is created, which is known as VMware. Malware is run in an environment where it is detected through creation, deletion, and new log entries; based on this running activity, the file is considered as a file that harms the system. Some files that are not examined properly from the static analysis can be taken for dynamic analysis.

Figure 9.9 shows in which phase dynamic and static tools will work in malware detection [14]. This paper suggests the different ransomware detection approach that analyzes the attack on the current running state of a computer by using the proposed model called finite state machine, which is used to synthesize the knowledge of the attack in the victim's system. Also, the author monitors the current system in terms of utilization, persistence through which the source of the attack can be discovered.

9.3   MATERIALS & METHODS

In this section, we see the various algorithms that are used for the prevention and detection of malware, as well as ransomware attack. Machine learning is the technique that is now trending in the detection of attacks, as we see in the previous section. In the testing phase, all the machine-learning algorithms are used in [5] SVM classifiers. Machine learning is not the only technique; there is deep learning [15], which is used for feature exploration in the Android system. The author proposes the TC-droid feature; it is a framework that depends on the text-classification method. TC-Droid implements the convolution neural network (CNN) to inspect the important information or knowledge. This extraction process has four types of features: permission, service, intent, and receiver. Once the feature extraction process is complete, the author trains the TextCNN for text classification. This is done in two categories, malicious and benign. We understand different attacks that happened and different techniques used for preventing attacks through Table 9.2.

In the above discussion, we saw the deep-learning and machine-learning techniques [2], but IoT is also one of the areas where we need to discuss ransomware

TABLE 9.2

Different Attacks Uses Different Techniques

Paper name	Types of Attack	Technique	Limitation/Future
Davies et al. [16]	1. NotPetya 2. Bad rabbit 3. Phobos hybrid ransomware	1. Recovery by using the AES Key from volatile type of memory helps in finding out the encrypted keys.	1. This technique is not implemented in all the ransomware families, like WannaCry.
Hwang et al. [17]	1. Ransomware attack	1. Markov model 2. Random forest model is a two-staged ransomware detection technique.	1. When ransomware detection has a higher false-negative rate, it is disastrous for the system. 2. More ransomware is required to generate like normalWare.
Vinayakumar et al. [18]	1. Ransomware attack	1. Two configurations of SVM: one is the linear kernel, and the second one is the radial basis function.	1. This technique uses complex architecture. An enhanced MLP network can be made in the future.
Arabo et al. [19]	1. Ransomware attack 2. Benign attack 3. Malware samples	1. Uses the interrelation between process behavior, as well as its nature, to determine the ransomware. This can be done by using machine-learning techniques with a low false-positive and false-negative to distinguish the benign and other ransomware.	1. In the future, the author will try to develop an early warning system that tells the victim in the first five seconds that a malicious attack has happened.
Mercaldo et al. [20]	1. Mobile Malware attack through image	1. By using the deep neural network architecture and to convert a binary to an image it can be done to encode a binary file into a PNG	1. In the future author tries to use the formal verification method for an image that creates better performance.
Lei et al. [21]	1. Ransomware attack	1. A self-recovery service is proposed by the author to ease the damage done by the ransomware attack to the edge servers. Evaluation is done based on a. CPU usage b. Disk Access Time	1. SRS is the solution for IoT systems and in the future more of this type of solution may be suggested.

(Continued)

TABLE 9.2 (Continued)
Different Attacks Uses Different Techniques

Paper name	Types of Attack	Technique	Limitation/Future
		c. Disk Read/write speed d. Disk Average transfer rate.	
Zimba et al. [22]	1. Wannacry Ransomware attack	1. A cascaded network segmentation approach to uncover the artifacts which set the security in the network. 2. Through static malware, the analysis author finds out the different techniques to propagate the ransomware and attack the CI components over the various network partitions.	1. This technique can be implemented in intrusion detection systems in each segment in the future.

because the internet of things is a network of the network where physical devices are connected. In IoT [23], physical devices can handle and organize the device; that's why they are called smart devices. This paper presents the crypto-wall attack works on detection model based on the communication, as well as the behavioral study of cryptowall for IoT. The suggested model works on the network traffic, especially TCP/IP traffic [24] over a server that extracts TCP/IP header and uses the C&C server for blacklisting to find out the ransomware attacks.

9.3.1 Algorithms Used for Ransomware/Malware Detection

The author proposed [25] a new technique that uses the machine-learning technique to detect and classify the spreading phase of a ransomware attack affecting the integrated clinical environment here. OC-SVM and Naïve Bayes detecting strategy work very well in detecting some of the dangerous malware like Wannacry, Petya, BadRabbit, and PowerGhost. The author also finds the worst case and best case of detecting the ransomware attack. Besides this, author uses the anomaly-detection method, and probabilistic classification technique of ML technique selection has taken place. Evaluation is done by using different measures like precision, recall, and accuracy, as well as F1-score through which the author analyzes the previous and proposed method improvement. Further, we understand the algorithmic point of view through the given in Table 9.3 below.

TABLE 9.3

Algorithm Used for Different Attacks

Paper Name	Attack and Techniques	Algorithm Used/Work done
Liu et al. [26]	1. Ransomware attacks and tracks the online behavior and understands each source, as well as destination entity.	1. LooCipher is used to detect the ransomware, which works on the approach called Identify-Collect-Examine-Analyze-Present.
Mitchell et al. [27]	1. North Korean Ransomware 2. Russian Ransomware 3. Chinese Ransomware	1. Various algorithms like AES for encryption techniques.
Zavarsky et al. [28]	1. Ransomware attack, and detection is done for the Windows operating system.	1. Uses the Message digest-5 for each sample for checksum values that are analyzed 2. PEiD tool is used to reveal packers, cryptors, and compilers
Alsoghyer et al. [29]	1. Android Ransomware attack and machine learning is used for detection and for doing permission analysis	1. The sequential minimal optimization algorithm 2. Naïve Bayes algorithm for cross-validation testing. 3. Random forest and Decision trees show the best result.
Sabharwal et al. [30]	1. Ransomware attack done by using phishing email, or spam or fake software up-gradation	1. An examination can be done by using the cloud infrastructure.
Saeed et al. [31]	1. Ransomware attack and detection is done by using machine-learning techniques	1. Feature extraction is done by generating the N-gram vector done by using an algorithm called CF-NCF (Class Frequency Non-Class Frequency) and train through the classification of ransomware, benign, malware.
Morato et al. [32]	1. Crypto ransomware attacks by monitoring the traffic over the network.	1. REDFISH algorithm is used for detection technique and also sequentially provides the events on a time basis.
Quinkert et al. [33]	1. Ransomware attack and technique is RAPTOR (Ransom Attack Predictor); it uses the domain classifier for the malicious domain prediction, and also does the time series forecasting; then it predicts ransomware domain.	1. Prediction is done using the Hidden Markov model (HMM model). 2. Another algorithm for prediction involves autoregressive models (ARIMA model) for forecasting events that include malicious domains. 3. For external signals, the algorithm used ARIMAX Model (autoregressive integrated moving average with exogenous variables)

(Continued)

TABLE 9.3 (Continued)
Algorithm Used for Different Attacks

Paper Name	Attack and Techniques	Algorithm Used/Work done
Kaur [34]	1. Ransomware attack prevention in IoT devices.	1. Blockchain, honeypot, and cloud & edge computing is the prevention technique used for IoT device.
Chakkaravarthy et al. [35]	1. Cyber Attack	1. Intrusion Detection Honeybot consist of a honey folder, Audit watch, and complex event processing

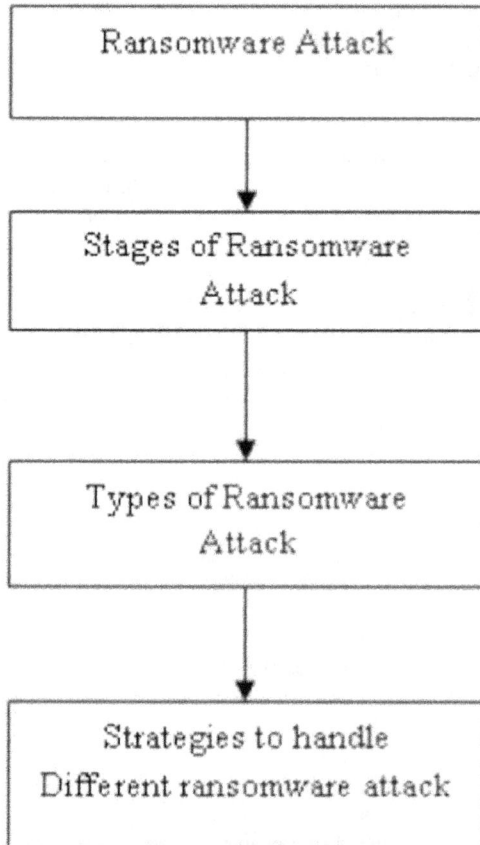

FIGURE 9.10   Architecture of this paper.

9.4 ANALYSIS

A data set is related to files that are either benign or normal; ransomware is used, and inside these various measures are works based on binary classifiers on true negative, true positive, false positive, false negative. Now, these four types are based on the actual value and predicted value. Figure 9.10 shows the architecture of this paper, which is based on ransomware attacks.

Precision: When fractions of positive no. of cases among the total no. of positive cases in the system are retrieved, then it is known as positive-class prediction, which is another name of precision. When it predicts yes, the person likes cats, how often is it correct?

Recall: No. of positive class predictions out of positive class in the data set. It answers the questions that when it is actually yes, the person likes cats, how often does it predict correctly?

F-measure: It is a measure that takes both precisions and recalls into account, and when one finds out the weighted average of precision and recalls, then it is known as F-measure; it also defines as a parameter that makes a compromise to be reached concerning precision and recall.

This is another perspective of handling the ransomware attack, that the victim fulfills the demand of the attacker and makes the payment, but it doesn't guarantee that future attacks won't happen from the same attacker. But then in some cases, it is useful. Now the question arises "how and when" [36] so the answer given by the author through semi-autonomous ideas in which strategy is implemented that payment is done only when the attacker gives the correct decryption key and uses the pay-per-file paradigm, still there is chance of loss of money because the attacker can deny any time whether the minimum requirement of money is done. Due to economic profit, it evolves among the criminals, and a survey by the author [37] says that increases the demand for ransom from \$294 to \$679 for 2014 to 2016.

9.5 PRACTICAL IMPLICATION

Through Discussion, we understand that antivirus software present for the system to prevent attacks, but as the hacking technique is evolving rapidly, this antivirus doesn't detect the attack, especially those attacks that come as Trojan and attack the system. At the time the victim knows that an attack has happened, it will be very late, and data loss will already have happened, which can be in different ways. Sometimes the attacker only wants to access the data; sometimes the attacker passes the information to the public platform to destroy the reputation of victims or organizations; sometimes the attacker demands money in exchange for data. In this paper, we discuss this type of attack only. So, there are various algorithms developed for detecting the ransomware attack, like RAPTOR, which is used to predict the attack by using the hidden Markow model, which distinguishes the normal files with ransomware files over the network. There is one more cybersecurity technique that works on the path of attack; if in any stage this path is interrupted, then the attack will not happen. In cybersecurity, many algorithms work on the concept of identify-collect-examine-analyze-present.

9.6 CONCLUSION

There are different threats for any system that are present over the network like attacks. All these attacks that we discussed above are those which are dangerous for the system and spread over the network. Firstly, we discuss malware; there are various types of malware present in the system to create damage in different ways; various malware works on the concept that is attached to the legitimate file or image through which the attacker attacks in the victim's system. Ransomware is the attack that makes the major damages; in terms of money, it creates a great loss for the victim. Ransomware has also different types like crypto and locker ransomware; crypto works on an encrypting concept, while locker works on the Android devices where it locks the screen and demands the money. There are further categorizations based on evolution over time, like crypto wall, BadRabbit, and many more, which we discussed in previous sections. All these evolution-based ransomware work on the demand of money in different ways, and the path associated with all this ransomware are different, but the base is the same, using the four steps that we discussed in Figure 9.2.

REFERENCES

[1] Almadhoor, Lama. "Social media and cybercrimes." *Turkish Journal of Computer and Mathematics Education (TURCOMAT),* vol. 12, no. 10, 2972–2981, 2021.

[2] Kok, S., Abdullah, A., Jhanjhi, N., and Supramaniam, M. "Ransomware, threat and detection techniques: A review." *International Journal of Computer Science and Network Security,* vol. 19, no. 2, 136, 2019.

[3] Tandon, Aditya, and Nayyar, Anand. "A comprehensive survey on ransomware attack: A growing havoc cyberthreat." *Data Management, Analytics and Innovation,* pp. 403–420, 2019.

[4] Chesti, Ikra Afzal, Humayun, Mamoona, Sama, Najm Us, and Jhanjhi, N. Z. "Evolution, mitigation, and prevention of ransomware." In 2020 2nd International Conference on Computer and Information Sciences (ICCIS), IEEE, pp. 1–6, 2020.

[5] Subedi, Kul Prasad, Budhathoki, Daya Ram, and Dasgupta, Dipankar. "Forensic analysis of ransomware families using static and dynamic analysis." In *2018 IEEE Security and Privacy Workshops (SPW)*, IEEE, pp. 180–185, 2018.

[6] Bijitha, C. V., Sukumaran, Rohit, and Nath, Hiran V. "A survey on ransomware detection techniques." In International Conference on Secure Knowledge Management in Artificial Intelligence Era, Springer, pp. 55–68, 2019.

[7] Fernando, Damien Warren, Komninos, Nikos, and Chen, Thomas. "A study on the evolution of ransomware detection using machine learning and deep learning techniques." *IoT,* vol. 1, no. 2, pp. 551–604, 2020.

[8] Kok, S. H., Abdullah, Azween, Jhanjhi, N. Z., and Supramaniam, Mahadevan. "Prevention of crypto-ransomware using a pre-encryption detection algorithm." *Computers* vol. 8, no. 4, 79, 2019.

[9] Hsiao, Shou-Ching, and Kao, Da-Yu. "The static analysis of WannaCry ransomware." In 2018 20th International Conference on Advanced Communication Technology (ICACT), IEEE, pp. 153–158, 2018.

[10] Bajpai, Pranshu, and Enbody, Richard. "Attacking key management in ransomware." *IT Professional,* vol. 22, no. 2, pp. 21–27, 2020.

[11] Aidan, Jagmeet Singh, Verma, Harsh Kumar, and Awasthi, Lalit Kumar. "Comprehensive survey on petya ransomware attack." In 2017 International Conference on Next Generation Computing and Information Systems (ICNGCIS), IEEE, pp. 122–125, 2017.

[12] Ren, Amos Lohj Yee, Liang, Chong Tze, Hyug, Im Jun, Broh, Sarfraz Nawaz, and Jhanjhi, N. Z.. "A three-level ransomware detection and prevention mechanism." *EAI Endorsed Transactions on Energy Web,* vol. 7, no. 27, 2020.

[13] Alayda, S. "Terrorism on dark web." *Turkish Journal of Computer and Mathematics Education (TURCOMAT),* vol. 12, no. 10, pp. 3000–3005.

[14] Kok, S. H., Abdullah, Azween, and Jhanjhi, N. Z.. "Early detection of crypto-ransomware using pre-encryption detection algorithm." *Journal of King Saud University-Computer and Information Sciences,* 2020.

[15] Hussain, Syed Jawad, Ahmed, Usman, Liaquat, Humera, Mir, Shiba, Jhanjhi, N. Z., and Humayun, Mamoona. "IMIAD: Intelligent malware identification for android platform." In 2019 International Conference on Computer and Information Sciences (ICCIS), IEEE, pp. 1–6, 2019.

[16] Davies, Simon R., Macfarlane, Richard, and Buchanan, William J. "Evaluation of live forensic techniques in ransomware attack mitigation." *Forensic Science International: Digital Investigation,* vol. 33, p. 300979, 2020.

[17] Hwang, Jinsoo, Kim, Jeankyung, Lee, Seunghwan, and Kim, Kichang. "Two-stage ransomware detection using dynamic analysis and machine learning techniques." *Wireless Personal Communications,* vol. 112, no. 4, pp. 2597–2609, 2020.

[18] Vinayakumar, R., Soman, K. P., Velan, K. K. Senthil, and Ganorkar, Shaunak. "Evaluating shallow and deep networks for ransomware detection and classification." In 2017 International Conference on Advances in Computing, Communications and Informatics (ICACCI), IEEE, pp.259–265, 2017.

[19] Arabo, Abdullahi, Dijoux, Remi, Poulain, Timothee, and Chevalier, Gregoire. "Detecting ransomware using process behavior analysis." *Procedia Computer Science,* vol. 168, pp. 289–296, 2020.

[20] Mercaldo, Francesco, and Santone, Antonella. "Deep learning for image-based mobile malware detection." *Journal of Computer Virology and Hacking Techniques,* pp.1–15, 2020.

[21] Lei, In-San, Tang, Su-Kit, Chao, Ion-Kun, and Tse, Rita. "Self-recovery service securing edge server in IoT network against ransomware attack." in *IoTBDS,* pp. 399–404, 2020.

[22] Humayun, Mamoona, Niazi, Mahmood, Jhanjhi, N. Z., Alshayeb, Mohammad, and Mahmood, Sajjad. "Cyber security threats and vulnerabilities: A systematic mapping study." *Arabian Journal for Science and Engineering,* vol. 45, no. 4, 3171–3189, 2020.

[23] Chatterjee, J. M., Kumar, R., Pattnaik, P. K., Solanki, V. K., and Zaman, N. "Preservação de privacidade em ambiente intensivo de dados." *Tourism & Management Studies,* vol. 14, no. 2, 72–79, 2018.

[24] Humayun, Mamoona, Jhanjhi, N. Z., Alsayat, Ahmed, and Ponnusamy, Vasaki. "Internet of things and ransomware: Evolution, mitigation and prevention." *Egyptian Informatics Journal* vol. 22, no. 1, 105–117, 2021.

[25] Fernandez Maimo, Lorenzo, Huertas Celdran, Alberto, Perales Gomez, Angel L., Garcia Clemente, Felix J., Weimer, James, and Lee, Insup. "Intelligent and dynamic ransomware spread detection and mitigation in integrated clinical environments." *Sensors,* vol. 19, no. 5, p. 1114, 2019.

[26] Liu, Te-Min, Kao, Da-Yu, and Chen, Yun-Ya. "LooCipher ransomware detection using lightweight packet characteristics." *Procedia Computer Science,* vol. 176, pp. 1677–1683, 2020.

[27] Almrezeq, Nourah. "Cyber security attacks and challenges in Saudi Arabia during COVID-19." *Turkish Journal of Computer and Mathematics Education (TURCOMAT)*, vol. 12, no. 10, 2982–2991, 2021.

[28] Zavarsky, Pavol, and Lindskog, Dale. "Experimental analysis of ransomware on windows and android platforms: Evolution and characterization." *Procedia Computer Science*, vol. 94, pp. 465–472, 2016.

[29] Alsoghyer, Samah, and Almomani, Iman. "On the effectiveness of application permissions for Android ransomware detection." In 2020 6th Conference on Data Science and Machine Learning Applications (CDMA), IEEE, pp. 94–99, 2020.

[30] Sabharwal, Simran, and Sharma, Shilpi. "Ransomware attack: India issues red alert." In *Emerging Technology in Modelling and Graphics*, Springer, pp. 471–484, 2020.

[31] Saeed, Soobia, Jhanjhi, N. Z., Naqvi, Mehmood, Humayun, Mamoona, and Ahmed, Shakeel. "Ransomware: A framework for security challenges in internet of things." In 2020 2nd International Conference on Computer and Information Sciences (ICCIS), IEEE, pp. 1–6, 2020.

[32] Morato, Daniel, Berrueta, Eduardo, Magaña, Eduardo, and Izal, Mikel. "Ransomware early detection by the analysis of file sharing traffic." *Journal of Network and computer Applications*, vol. 124, pp. 14–32, 2018.

[33] Quinkert, Florian, Holz, Thorsten, Hossain, K. S. M., Ferrara, Emilio, and Lerman, Kristina. "RAPTOR: Ransomware attack predictor." arXiv preprint arXiv:1803.01598, 2018.

[34] Kaur, Jaspreet. "A secure and smart framework for preventing ransomware attack." arXiv preprint arXiv:2001.07179, 2020.

[35] Chakkaravarthy, S. Sibi, Sangeetha, D., Cruz, Meenalosini Vimal, Vaidehi, V., and Raman, Balasubramanian. "Design of intrusion detection honeypot using social leopard algorithm to detect IoT ransomware attacks." *IEEE Access*, vol. 8, pp. 169944–169956, 2020.

[36] Delgado-Mohatar, O., Sierra-Cámara, J. M., & Anguiano, E. "Blockchain-based semi-autonomous ransomware." *Future Generation Computer Systems*, vol. 112, pp. 589–603, 2020.

[37] Cartwright, Anna, Hernandez-Castro, Julio, and Cartwright, Edward. "An economic analysis of ransomware and its welfare consequences." *The Royal Society*, 2020.

10 Security Management System (SMS)

*Shahida, Khalid Hussain Usmani,
and Mamoona Humayun*

CONTENTS

DOI: 10.1201/9780367808228-10

10.1 INTRODUCTION

This is an era of ubiquitous computing, and digital devices have become imperative for businesses and institutions, which leads to information that is ineludible in all kinds of systems, industries, and organizational activities; therefore, protecting information as assets is necessary. This must be done against risk of deprivation, outrage, destruction, exposure, and mistreat of information [1]. In the sense of digital technology trends, information is raising without bounds, and every type of business needs competitive tools to build security systems to prevent information leakage to the outside world and protect the entire society of the business world [2]. There is a necessity to build a strong security structure, which should be effective for the organization's security system. In the view of all these requirements, an appropriate system is designed to fulfill the needs of the digital environment to secure data and information [3], which is known as the security management system (SMS). In this section, we will revise the elementary concepts of information security, discuss the security management system for the information specifically, and discuss how the associated risk could be tackled utilizing security management system techniques. All the critics and complexities included in this regard will be explained further.

10.2 APPLICATION OF INFORMATION SECURITY IN CORPORATIONS

Information security is the way of safety measures and reasonably protecting the information by qualifying the information risks for all types of corporations, from small to medium enterprises all out to the bigger corporate level. It, in fact, has been a bigger challenge to many organizations. Moreover, information security became an existent risk management process that surrounds all the information that needs to be preserved [4]. Information security and computer security terms are mistakenly applied interchangeably, but also are interrelated in the form of common goals of securing information from the outside world employing confidentiality, integration, and accessibility. Hence, there is a slight difference among them, such as information security related to the confidentiality, integrity, and availability (CIA) of information, irrespective of its existing forms, including prints and electronics (as mentioned in Figure 10.1). The computer security basis to ensure the convenient functioning and accessing of a computer system is irrelevant of the present information processing. Although information security is all about the clean access and protection of sensitive data, additionally, other features, for instance, legitimacy, quick responsiveness, reliability, validity, and the definitive and certainty of information can also classify in protection.

10.2.1 Information Security Components: Confidentiality, Integrity, Availability (CIA)

It has been discovered that modern organizations have not complied with the sound governance principles of information security, specifically, acquiescence with the

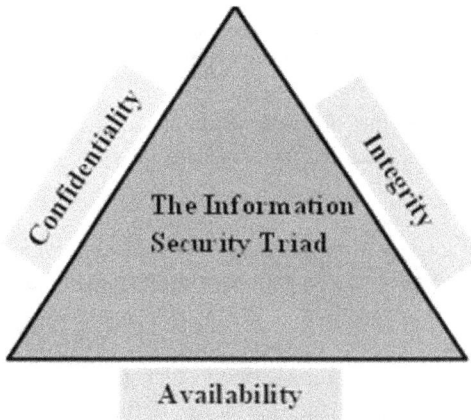

FIGURE 10.1 The CIA group [5].

information safety and monitoring policies, and that is mainly due to the confined resources and expertise [5]. Keeping the sensitive information secured, an organization needs to set up prerequisites potentially that aim to guide information security policies within a business.

10.2.1.1 Confidentiality

In connection with information security, the term *confidentiality* relates to the set of rules, which act to limit the information access [5]. It ensures that only authorized information and authorized people are allowed to approach the information and prohibition of possession from the unauthorized people. In other words, we consider the confidentiality as privacy. Confidentiality is achieved through the methods, such as user ids and passwords, access control lists, and policy-based securities [6]. In online banking, an account number is taken as an example for confidentiality method. Also, in networks, data encryption is a standard method to certify confidentiality. Although user ids and passwords are accounts for typical procedures, two-factor authentication process is playing a role for discretion [7]. Some other procedures do share some cautious actions, including biometric verification and software tokens.

10.2.1.2 Integrity

Integrity works out to assure that the information has not been altered, while legally retrieved, and intended use is truly observed. It aims to certify that the information format is not tampered with and is being perceived as the creator proposed the user to have [8]. Integrity is related to the reliability of the information, even though stored in the causal systems, warehouses and databases, etc. It must be reformed by an acknowledged process and preserved thorough file permissions and access controls. Some security mechanisms are used to execute the integrity in an organization; theseh include data encryption and hashing. Many tools implement the 'one-way hash'. For instance, when data is to be transited to the receiver, a hash of specific data is also calculated and moved along with the original message/data. When the receiver receives the data, the hash message is compared with the hash received [5]. If both are the same, the data is in the original format; otherwise, it has lost its value. Sometimes data is modified by the nonhuman-caused events, just like server crashes. To avoid these changes and recover the originality, there must be redundant systems and backup procedures already observed, so that data integrity could be ensured.

10.2.1.3   Availability

Any organization's efficiency depends upon its services, which should be readily available for its authorized people. Availability in CIA triad makes sure that the data and information is readily and smoothly available when required, and in appropriate time-frame [5]. This timeframe means different things for different companies/industries; some companies could not bear a crash or the unavailability of even one second in 24 hours a day, and it would be fine for some if their site suffered some minutes occasionally. Availability is employed using some approaches, for example, hardware maintenance, network optimization, and software patching/upgrading. Sometimes, it is obvious that hardware issues occur; in such conditions, RIAD [7], failover, redundancy, and high availability clusters are involved to alleviate the significant out-turn. Distributed denial of services (DDOS) attacks [9] are malicious actions designed to bring down the respective services and thus overthrow availability [10]; in such conditions, hardware systems can protect against downtime and inaccessible data. Any natural or man-made disaster can cause the availability to be thrashed. For the said case, usually companies try to design fault-tolerant systems, which includes redundant systems and drives, etc.

Some variety of tools should be available to certify the confidentiality, integrity, and availability of material in the organization. So, the information security procedure could be observed utilizing these tools. For example, authentication, access control, encryption, backups, firewalls, and virtual private network, etc. [5].

10.3   INFORMATION SECURITY INCIDENTS [11]

1. Vulnerability. It is a faintness of computer devices, which could be exploited by threat attacker. Or sometimes it refers to a flaw in the system that could provide a room for attack and exposition of information.
2. Threats. These are unwanted events that potentially do harm to a computer system, often resulting in serious damage. Threats can cause destruction to the computer system, networks, and more.

10.4   COMMON ATTACKS ON CIA TRIAD [12]

1. Attacks threatening Confidentiality. Threat is one of the information security incidents, and there are generally two types of the attack that could threaten the information confidentiality:
   a.   Snooping
   b.   Traffic Analysis

   Snooping denotes the unauthorized access to, or interception of, information (eavesdropping), whereas traffic analysis introduces the types of information that an invader monitors by online traffic.

2. Attacks threatening Integrity. Some known attacks that cause information to be at risk are modification, masquerading, replaying, and repudiation.
3. Attacks threatening Availability. For the sake of unavailability of information, denial of services (DOSs) attacks may decelerate or suspend the required services. This could result in making the system continuously busy and collapsing the system so that in two-way communication, a sender sometimes believes that a sent message didn't deliver to the receiver and must be resent.

10.5 INFORMATION SECURITY MANAGEMENT

Information security management contributes in the organization of information to set the required goals, provide secure information, and assess every upfront obstruction to those goals and render appropriate solutions for further actions [13]. Management of information is a part of information security limits, as elaborated in the Figure 10.2.

10.6 SECURITY MANAGEMENT SYSTEM

Security management system (SMS), sometimes known as information security management system (ISMS), is generally a section in an establishment that imparts management strategies built on risk management and risk assessment. The target is to originate, deploy, utilize, administrate, sustain, and boost the information security [15]. It is categorized as the systematic and quantifiable

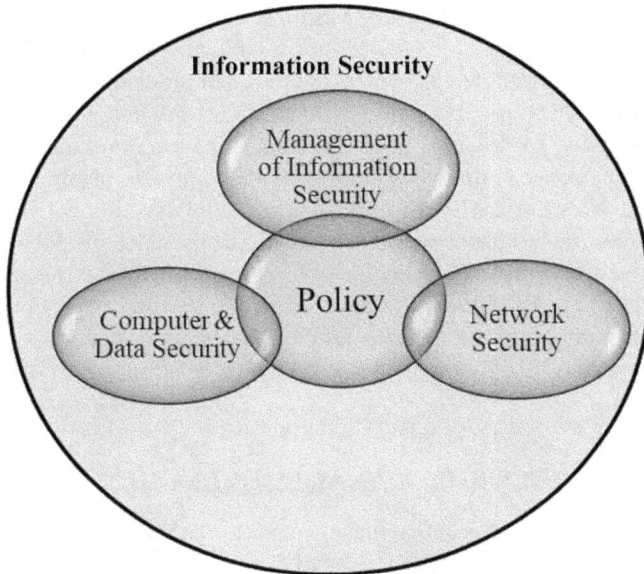

FIGURE 10.2 Main components of information security [14].

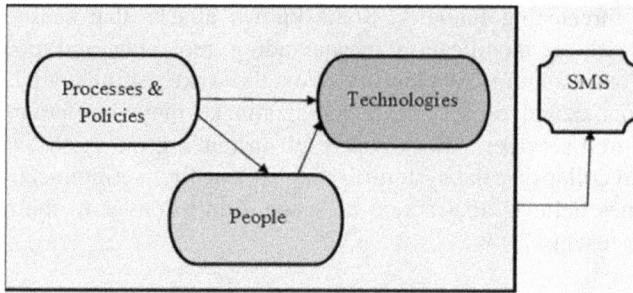

FIGURE 10.3 SMS participants [3].

approach that smooths the sensitive information for the sake of protection. In management systems, information and security are not only related to configure a simple firewall or confine a contract with security providers. Following that approach, it is necessary to keep balance with the security activities and adhere to the common tactics to facilitate the optimization in the reasonable protection level [16].

Conforming to the National Institute of Standards and Technology (NIST) [17], today's SMS is mainly structured on complex basis that includes technologies (i.e., software, hardware, and firmware), policies and procedures, and the people working in that enterprise to furnish that business with the ability to control, operate, store, retrieve, and timely spread in a way to enhance the business functionality, as shown in Figure 10.3.

Security management system, in fact, relates to the collection of procedures scrutinized with information technology (IT) type hazards or sometimes information security management (ISM). The SMS must have all permissions to access and retrieve system information of the organization, following the ISO 27001 [3], which is a standard security framework, security tactics, and proclamation for information security controls. The organization's business could go into harm control, risky situations, possibility of financial loss, and legal obligation in the absence of these strong security controls. Employees in any firm have private or public records that need to be safe and secure. Also, this information could exist in many forms, and all have some values toward the concerned entity. In the presence of globally occurring information threats to organizations, the enforcement of SMS is inevitable. In literature it is stated as [18] "it is a life-cycle approach to impose, sustain and further bring improvements to the set of schemes, standards, and modes that is concerned with the information assets of organization require to be in a feasible manner for strategic intents."

10.7 BASIS OF SECURITY MANAGEMENT SYSTEM

Two major mechanisms of security management system are risk management and risk assessment. These are comprehensive, known terms in literature. Here an affiliated view of risk management and risk assessment, mainly considered in terms of SMS as presented.

10.7.1 RISK MANAGEMENT

"The process of recognizing vulnerabilities and intimidations within the limits of an organization and assembles computations to try to minimize the affects over the informational resources as well" normally be known as risk management. Organizational assets are being protected and fortified by a combination of a few activities contained in risk management that intend to fulfill the missions and objectives of the organization; moreover, at the same time, they are intended to stabilize the management effort consumed on potential threats and attacks occurring on a probability evaluation. It is a persistent activity basically focused on the analysis, setups, treat exposures, control, and monitoring of deployed measurements, along with the enforced security policies (shown in Figure 10.4).

Risk analysis has the major role in management to process the influenced factors in the context of information security. When identification and estimation procedures are successfully applied on the assets, then risk management and alleviation would analyze the issues of information security incidents such as threats and vulnerabilities. Afterward, the mitigation is the proposed method to process and minimize the influence and likelihood of these incidents. Risk management process contains some processes explicated below:

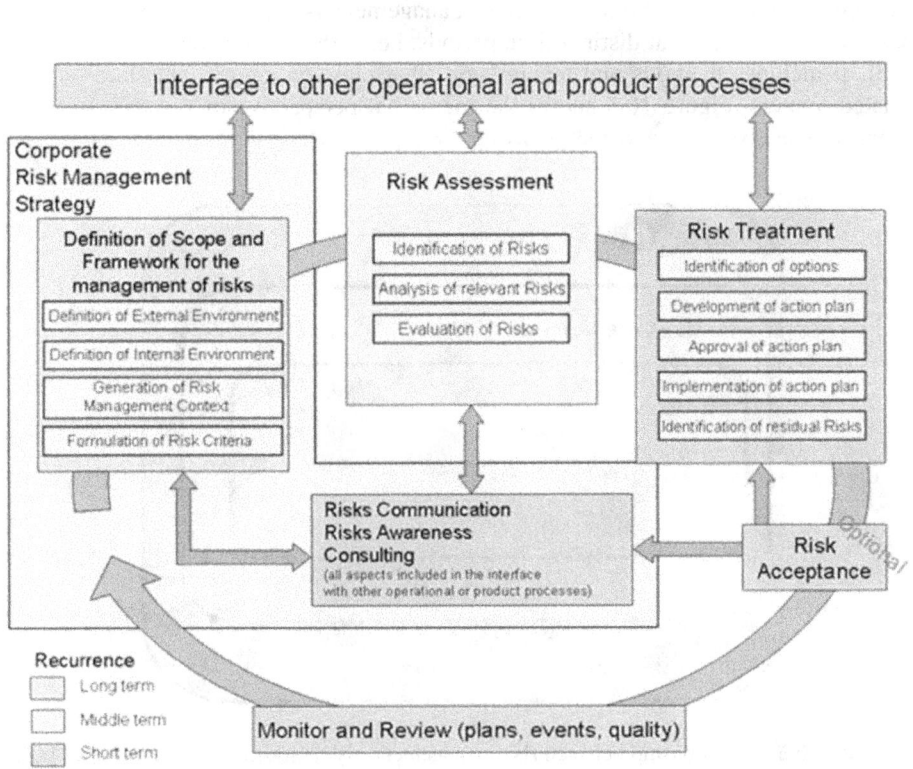

FIGURE 10.4 Risk management process [14].

1. Risk Assessment: with risk identification, risk analysis and risk valuation
2. Risk Treatment: to first discover the risks and later apply measures to modify risks
3. Regular Monitoring and Review: to ensure the effectiveness of risk management
4. Risks Based Communication: to awareness and exchange information about risks to internal and external stakeholders
5. Risk Acceptance: done by the high-level management either to reduce, off-load, accept, or ignore by viewing the feasibility

10.7.2 RISK ASSESSMENT

Risk assessment has main upshots:

1. It governs the threats.
2. Threats have some risky aspects, which could be prioritized by the assessment.
3. Expound the controls and utilize the protection counts.
4. Afterward, develop a progress plan for the measures' execution.

Risk assessment is accepted as the part of the risk management process. But some discrepancies occur in both since risk management is frequent whereas risk assessment is scheduled at distinct time periods, i.e., either once a year or on demand [15]. It includes methods and technologies that classify, gauge, and state on risk related matters. Figure 10.5 shows the hierarchal perspective of the risk management, risk assessment, and risk analysis.

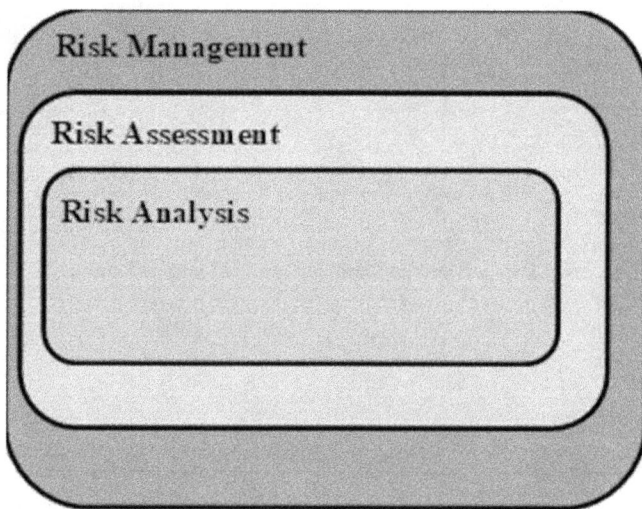

FIGURE 10.5   Relationship between risk management, risk assessment & risk analysis [18].

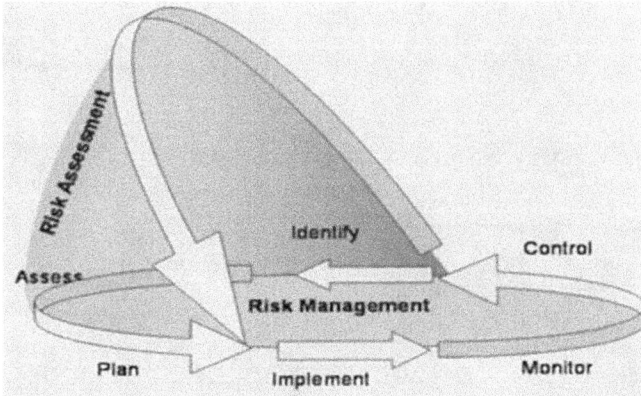

FIGURE 10.6   Risk management & assessment as dependent processes [19].

Risk assessment and risk analysis are being considered the subpart of risk management, so their technical relationship view is drawn in Figure 10.6.

It is worth declaring that risk management and risk assessment (as shown in Figure 10.4) are dependent processes that possess arrangements of activities.

10.7.3 CHALLENGES FOR RISK ASSESSMENT

There are some challenges may be confronted by information security risk assessment including [20]:

A.14.2.1. Lack of executive management assurance and support
A.14.2.2. Lack of suitable procedures for information security risk management
A.14.2.3. Efforts of all security staff not assembled
A.14.2.4. Inappropriate evaluation management
A.14.2.5. Assets rights either indeterminate or unrehearsed
A.14.2.6. Prevailing automated solutions are not fully supported
A.14.2.7. Existent various IT risk assessment schemes

10.8   SOME OTHER ESSENTIALS FOR SMS

In the presence of these two major components (i.e., risk management and risk assessment), SMS also has some other essentials [21], as follows:

1. Total quality management (TQM): Although the organization's various management systems are derived from the total quality management, it is related to the production environment and organization's security. Moreover, these two are interdependent in the terms of success for organization security. The application of TQM is purely set up on criterion of expertise and efficacy.
2. A tracking and declaration model built on abstraction layers

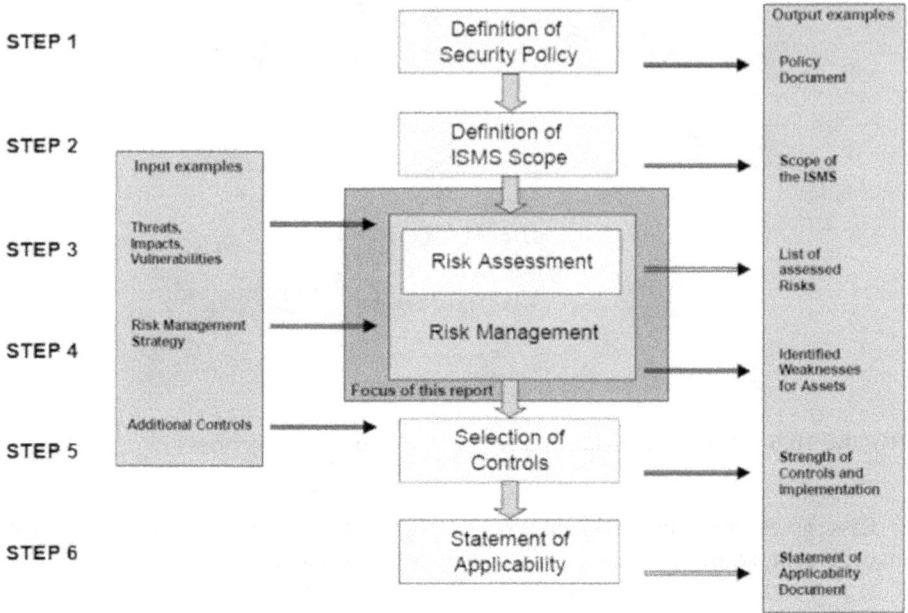

FIGURE 10.7 Security Management System emerging process [22].

3. A structures method: which encompasses individuals, procedures (methods), and related technology
4. A compliance agenda (framework): through which organization used to manage the information security acquiescence (shown in Figure 10.7)

10.9 STEPS INVOLVED IN THE INFRASTRUCTURES OF THE SMS

As information security is the constitutive part of any organization's operating environment and commercial culture; following the same concept, there are various steps that cover the building of SMS. There are some things to be taken into consideration while performing these steps [3]; first is to take input from all the concerned stakeholders in the organization and afterward discuss the results to follow the agreed-upon pathway. Central repository for SMS should be maintained and updated, which could be done through a security manual. It is the confidential and reliable document usually sustained by the chief security officer. Let's analyze the several steps incorporated in the establishment of SMS [23].

1. **The first step** starts with the definition of the scope of the SMS in the context of the aspects of the business, the firm, its place, present resources, and expertise about technology. The scope is based on the association of any interface with other system, organization, or any supplier. It also responds to security requirements satisfied by SMS.

2. **The second step** starts with the orderly approach of the risk assessment. Here risk assessment method for an SMS is identified, with the acknowledged business security and other legal requirements. The goal is to set policies and objectives for identification of the threats and vulnerabilities and pay attention to control and reduce the risks to some extent. By going along to the above procedure, we can, now, determine the risks criteria and classify the acceptable level of risk by management approval by identification of the impacts of confidentiality, integrity, and availability lacks on the organization's resources.

3. **The third level** is all about the top-down approach because the security of the information is a management issue, not only IT issue. Top management of the organization must play a critical role and take an ownership in the maintenance of SMS. It is the core responsibility of the management to motivate the other employees in the enterprise to adhere to security principles.

4. **The fourth level** is the definition of the functional roles. Once the management decides to act on the principles of the SMS, all functional roles must be definite. The type of the role to be assigned to any entity depends upon the type and magnitude of the firm, and then roles could be varied. An owner of the SMS who serves as chief information security officer must be appointed. Other functional roles, including data agents, security awareness mentors, etc., must also be assigned.

5. **The fifth step** is to outline the policy in the same terms as we defined the scope of the SMS, but it includes the framework for setting its aims, which clears the vision of information security direction and principles. The security management stratagem is a type of document that elaborates the enterprise's information security plans at a high level. Here, some regulatory requirements also be noted, requirements that are derived from the risk assessment. In this stage, a criterion is established to handle the risk involved, evaluate the risk, and define assessment, and that criterion must be affiliated with the business significances and objectives.

6. **The sixth step** is to write the standards. These are the fixed necessities that an organization must have to follow to support the security policy and its measurement. The best approach is to document what the consequences are and where we lack behind.

7. **The seventh step** is to find out and evaluate choices for the treatment of risks, after we have set the policies and standards. These are the useful actions, such as establishing proper controls, accepting risks empirically, providing the way to fully follow the organization's policy and criteria for accepting threats, and then avoiding risks.

8. **The eight step** is to select the control target and other oversights for dealing with the risks. One thing that must be under consideration is to choose the control that will be cost effective. For instance, the cost of the treatment must not exceed the cost of the impact of the risks that are planned to be reduced. Risks are not always of financial type; they could be related to safety, legal and personal information, regulatory compulsion, and reputations. So, these are effects must be considered.

9. **The ninth step** is to guide the procedure and prepare the statement of implication. The appropriate guidelines for the mitigations of the risks and threats should be followed in order to meet the standards of the policy. These procedures normally chosen by the people who device the control. Then, in the statement of implication, the document should be prepared to record the control objectives, and then control selection over terms and conditions.

10. **The last step** is also one of the core actions to be taken in SMS. In this step, the management approval is compulsory for the applicability of proposed residual risks control and operation in SMS.

The implementation phase [13]:

1. In order to import and settle the risk plan, review is inevitable.
2. Then implicate the risk handling and controls.
3. Review the related activities.
4. Properly monitor the implementation process.
5. Efficiencies and effectiveness should be checked out on regular basis.
6. If there are targeted risks, monitor them on priority basis.
7. Audits must be consistently guided.

The action phase [13]:

1. To update and carry out the further improvements
2. To choose appropriate convenient practices
3. To make sure the achievements of targets, maintain and progress where lack

As SMS facilitates with the core requirements for setting up the proper management system, so there are numerous standards and actions to implement for the formation and improvements of these activities, such as establishing, configuring, adaptation, restructuring, etc. Practically, many organizations adopt to initiate these methods, which are most suitable to the firm or business sector, and play the role of strategic decisions for an organization being scaled based on needs. For the said purpose, national or international standards are put into practice with the adaption of security mechanism, policies, and substructure. By accounting the above strategies, new better practices for a specific business are generated, and it can be put into practice by internal and external parties described by ISO/IEC 27000. It offers a set of controls that can be employed for SMS [24]. Here we will discuss the ISO/IEC 27001 following the mechanism of Plan – Do – Check – Act (PDCA) [20].

It is convenient to take a look some illustrations of control objective and system development control and preservation and business continuity management [25].

10.10   SYSTEM DEVELOPMENT AND MAINTENANCE

- System development and maintenance is about applying the security requirements for operating and application levels. Here, the control objective is to warrant that information security has been kept by avoiding the damage in terms of loss and preventing the amendment in the data further verify the misuse in user level. Some types of controls that could be applied; they include information security requirements, investigation and description in the business requirements with updates, input data authentication, internal processing controls that validate to detect the corruption in the processed data, and output data authentication to ensure that the given results are according to input data and stored information processing or not.

- Cryptographic controls, which aim to guard the confidentiality, authenticity, and integrity in the information security. In cryptographic, controls that could be applied are procedures and management in which a system must be executed based on the set of standards, management, and approaches with cryptographic techniques, encryption to protect the sensitivity of critical information, and non-repudiation services for the ensuring the occurrences of events or actions.
- System files security is the control objective to make sure that IT related activities are being carried out in secure manner. It applies the controls, for instance, control of operational software for production system and control access to program source library and database.
- Protection in growth and provision of processes, the control objective is to keep the safety of application software system and information. Controls applied are change control measures, technical analysis of operating system changes control that application system should be reviewed and verified when updates happen, and outsourced software development control.

10.11 BUSINESS CONTINUITY MANAGEMENT

Taking the view of business continuity management, the control objective is to confront disruption to corporate functionalities and handle business critics from the core catastrophes and failures. It gives the tolerance to the business defects and manages the business continuity. Implemented controls are:

- First control is the process of business continuity management to ensure the business maintenance overall in the organization.
- Second control is the business continuity and impact analysis to accomplish the risk assessment, strategy planning to approach the business continuity.
- Third control is all about the planning framework for the business continuity. In which a framework is structured out to make sure the consistent planning for an organization.
- Fourth control is to evaluate, maintain and re-assess the business continuity plans. Testing and maintenance will be carrying out according to the reviews to confirm that the effectiveness and latest trends are being done.

10.12 SECURITY MANAGEMENT SYSTEM COMPONENTS

Security Management System involves some critical components as expressed in Figure 10.8.

FIGURE 10.8 SMS components [26].

1. Management principles
2. Resources
3. Personnel
4. Information security process

10.13   SECURITY MANAGEMENT SYSTEM DOMAINS

The SMS standards include some security areas and control intentions. The domains and control goals for SMS are listed below in the Table 10.1.

10.13.1   ISO/IEC 27001

International organization for standardization (ISO) and international electro-technical commission (IEC) published some security standards known as ISO/IEC 27000 standards. ISO/IEC 27001 is one of the information security standards that is completely re-written and published in 2013 [28]. Herewith the ISO 27001 famil-iarized the complete set of security techniques along with the information security management system and other control requirements. This release of the standard formally specialized the security management system as other standards specify the different types of management systems in the ISO 27000 family standards [29]. The ISO/IEC 27001 industrialized to manage the information resources of an organi-zation and life cycle of the business [30]. It works under the management control and the organization, which tends to fulfill the all required steps to establish the security management system might be certified by the ascribed certification channel ensuring the successful completion of an SMS inspection [23].

TABLE 10.1
SMS Domain [27]

Domain	Objectives
Security Policy	In SMS we setup the policies to facilitate the management direction and support for information safety.
Organizational Security	It deals with the management of the information within the enterprise.
Asset Management	Attain and preserve the proper protection of organizational assets.
Human Resources	It concerned with the particulars of employees within the organization such as training, tasks, and employees' attitude towards the security occurrences.
Physical and Environmental Security	Avoid illegal physical assets approach
Communication and Operations Management	This element in the domain of SMS guarantees the precise and safe operations of information processing conveniences.
Access Control	Manageable admittance to information
Information System Acquisition, Development & Maintenance	Certify that the security is a mandatory part of the information management system.
Information Security Incident Management	In order to take some feasible actions, information security system measures are communicated.
Business Continuity Management	Management of business actions through critical conditions and tackle all types of issues.
Compliance	Ensure the follow up of information security necessities.

10.13.1.1   History of ISO/IEC 27001

BS 7799 was a standard formally published by BSI Group in 1995. It was inscribed by the United Kingdom government's Department of Trade and Industry (DTI), and it included numerous parts. Then the first part, which was all about the information security management, was then rewritten in the 1998 [30]. After some popularity in the worldwide discussion, ISO adopted the standard and discovered the ISO/IEC 17799, about "Information Technology – Code of practices for Security Management for information" in 2000. ISO/IEC 17799 was then reviewed in June 2005 and become as standard of ISO/IEC 27002. The subsequent segment of BS7799 was published in 1999, titled "Information Security Management Systems – Specification with guidance for use." It was purely about the implementation and controlling of security management system (SMS) for sensitive information of the enterprise [31]. This, afterward, became the ISO/IEC 27001 in November 2005. Also, the third of the BS7799 was merged and aligned with the ISO 27001, which covered the risk analysis, assessment, and management [11].

The international Organization for Standardization (ISO) has introduced some standards to secure and authorized access of the organization's information and information security management systems (ISMS), so included standard ISO/IEC 27001, which fulfills the criteria for "Information Technology, Security Techniques, Information Security Management Systems, Control Requirements" [32].

ISO/IEC 27001 proposed and worked on the phenomenon of 'Plan-Do-Check-Act' (PDCA) model, which intended to organize, execute, supervise, and continuously refine the efficiency of an organization's SMS [33].

10.13.1.2 PDCA Model

The management system for the information security process directs the establishment and maintenance of an organization by executing the SMS [29] and builds on the scheme of the Plan-Do-Check-Act (PDCA) model. The phases of the PDCA elaborated as under (Figure 10.9):

1. **Plan** – devise SMS policy, objectives, operations, and actions to control and minimize risks and improve information security to produce outcomes according to the enterprise's undertaking and goals.
2. **Do** – execute and function the policy of SMS, supervisions, processes, and procedures, attending the better practice approach.
3. **Check** – evaluate and compute the conduction of the process in contrast to SMS policy, objectives, and applied practice, and then report the outcomes to the high-level management for review and further actions.
4. **Act** – after the management reviewed and internal audit or other checking of the policy, objectives, and performance, take some preventive and corrective actions. These actions should be done to achieve frequent improvements of the SMS.

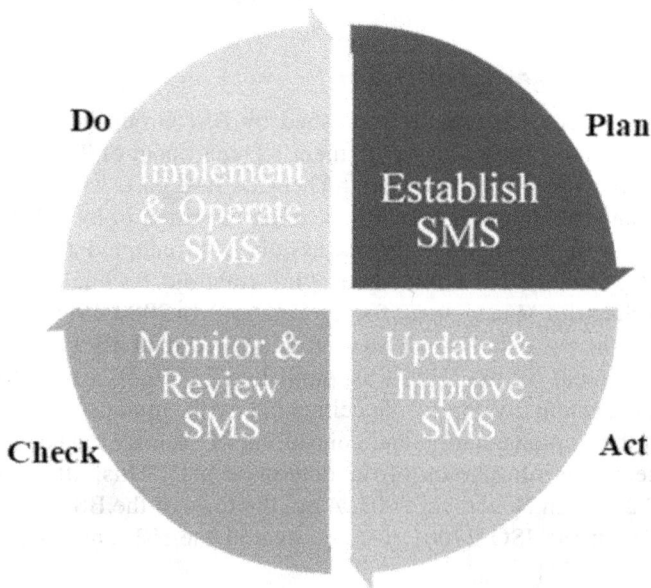

FIGURE 10.9   SMS Plan-Do-Check-Act cycle [20].

10.13.1.3   Changes in ISO/IEC 27001 in 2013 Version

As ISO/IEC 27001 has been fully re-written as a modified final version, as compared to the ISO 27001: 20015 [34]. It is a completely modified structure of the standard and puts more focus on computing and evaluating the SMS of an organization performance wise. A new section known as outsourcing was introduced, which means that numerous organizations may count on third parties to furnish them with some facets of information technology. It utilizes the PDCA but does not emphasize the implementation as before [30]. This standard version is more intended for the organizational framework of information security, and the criteria for risk analysis, risk assessment, and risk management has been rehabilitated. If summarizing the design for 27001:2013, it fits better along with other administration standards, for instance, ISO 9000 and ISO/IEC 20000, and is somewhat additional upgraded.

The structure of the standard contains a short clause (short clauses are covered earlier in the steps of the SMS) and a list of annexures that consists of controls and their objectives. Some new controls other than traditional have been added to the ISO/IEC 27001: 2013.

New *Controls* [30]:

A.6.1.5.   Information security in project management
A.12.6.2. Restrictions on software installation
A.14.2.1. Secure development policy
A.14.2.5. Secure system engineering principles
A.14.2.6. Secure development environment
A.14.2.8. System security testing
A.15.1.1. Information security policy for supplier relationships
A.15.1.3. Information and communication technology supply chain
A.16.1.4. Assessment of and decision on information security events
A.16.1.5. Response to information security incidents
A.17.2.1. Availability of information processing facilities

10.14   SECURITY CONTROLS [35]

Security controls are precautions and countermeasures to evade, spotify, counteract, or lessen the information security risks of any organization. These could be classified by following some criterion, for instance, according to the security incident:

1. Before the incident, *preventive controls* are planned to prevent the event from happening, i.e., by catching out unauthorized intruders
2. During the incident, *detective controls* are aimed to recognize and depict an occasion in progress, e.g., while encountering the burglar, activate the security of the system
3. After the incident, *corrective controls* are proposed to minimize the level of any damage originated by the event, i.e., retrieving the organization to normal functioning level as effectively as possible.

In the structure, it has been defined that how associations can react to risks encountered with the risk management and maintenance by selecting a proper control. A new feasible change has been made in 27001:2013 that Annex A is not required any more for handling the information security risks. This means that in the new version, Annex A is not acting as a proper control set. This empowers the risk evaluation, making it more manageable and more interesting toward the organization and building a proprietorship of both the risks and controls. This was the key change made in the newer version of SMS (ISO 27001). There are now 114 controls in 14 clauses and 35 control classes; whereas in 27001:2005, there were 133 controls in 11 types [35]. Given below are the number of controls for each clause in the new standard.

A.5. Information security policies (2 controls)

A.6. Organization of information security (7 controls)

A.7. Human resource security – (6 controls that are applied before, during, or after employment)

A.8. Asset management (10 controls)

A.9. Access control (14 controls)

A.10. Cryptography (2 controls)

A.11. Physical and environmental security (15 controls)

A.12. Operations security (14 controls)

A.13. Communications security (7 controls)

A.14. System acquisition, development and maintenance (13 controls)

A.15. Supplier relationships (5 controls)

A.16. Information security incident management (7 controls)

A.17. Information security aspects of business continuity management (4 controls)

A.18. Compliance; with internal requirements, such as policies, and with external requirements, such as laws (8 controls)

10.15 CERTIFICATION

No standard could guarantee to ensure the 100% working and efficiency of the applied system in the organization, but ISO 27000 has various standards to choose which an SMS can be certified [36]. An organization's SMS might be got certified, while acquiescent with the ISO/IEC 27001 by many accredited registrars worldwide. In different countries, the infrastructure that conforms with the management systems' standard formalities are called "certification bodies," in some places, it called "registration bodies," "assessment and registration bodies," and often "registrars" [24]. The ISO/IEC 27001 certification process encompasses a few stages external audit procedures defined by ISO/IEC 17021 and ISO/IEC 27006 standards.

The overall worldwide certified organizations for ISO/IEC 27001 are now 40,000 as increased by 20% annually (as shown in Figure 10.10). Interesting to know that the information technology division dominated the certification list, with 40% of certified establishments being in that commercial zone (elaborated in Figure 10.11 as per the survey of 2015).

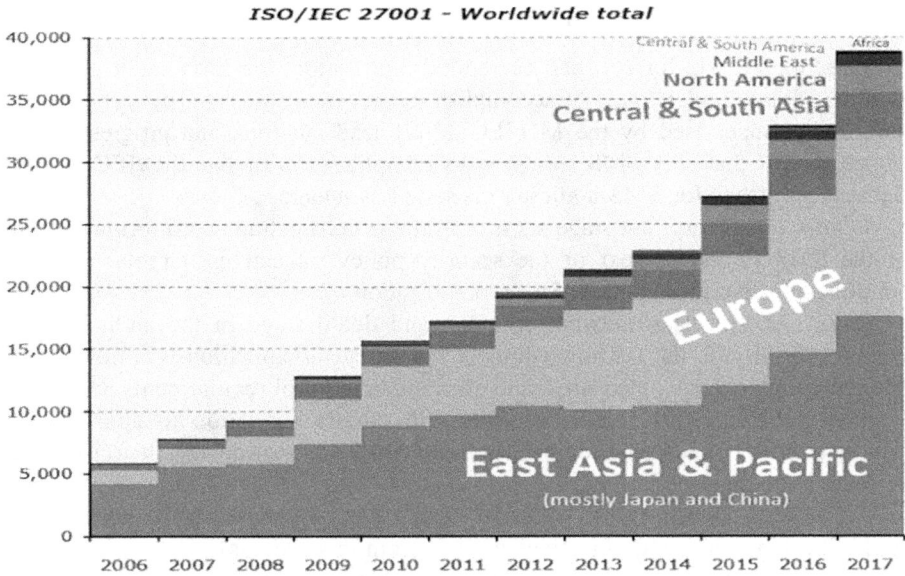

FIGURE 10.10   ISO/IEC 27001 certifications worldwide [31].

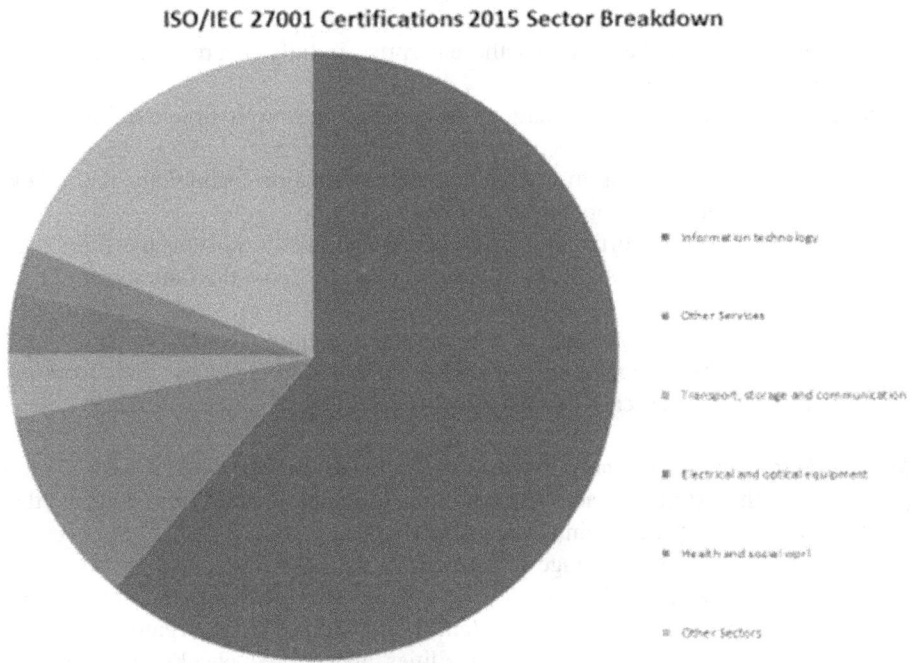

FIGURE 10.11   ISO/IEC 27001 certifications sectors [37].

10.16   SMS (ISO/IEC 27001) AUDITS [23]

To meet the criteria for certified accredited registration, the SMS audit could be scrutinized and implemented to accomplish the necessary stages. This certification audit being supervised by the ISO/IEC 27001 lead auditors, and progressing all stages certify that this SMS deserves to compliance with the ISO/IEC 27001. Subsequent stages for SMS audit are discussed as under:

Stage – 1: the very first stage starts with the understanding and informal review of the SMS in the context of the security policy, objectives, targets, required controls, and risk management for an organization.

Stage – 2: this is somewhat complex and detail stage in the audit. As we have to go through the acknowledgment that either organization is following its defined policies, goals, measures, and all standard control requirements according to the ISO/IEC 27001. Later, auditors will pursue indication to approve the risk assessment of information security and SMS out-turn design. It trails some inspected sub-stages:

1. Checking aims and objectives derived from this process
2. Checking performance monitoring, analyzing, reporting, and reviewing according to the stated targets.
3. Checking assigned responsibility and management of the high and sub-ordinates management level for the security, validation of the checklist that proper observance has been reviewed, and a follow up is made for infrequent actions.
4. Checking and make sure that the enterprise is fully aware of the IT infrastructure significance.
5. Checking the underlying matters, to realize that how is organization is equipped:
   • Does the organization certify that the information technology is the key component of the organization plans?
   • Is the IT sector officially systematized and able to sustain the fully segregation of responsibilities in meanwhile to facilitate the functions of IT in the organization?
   • Is the organization support to retain the key employees?
   • How is the organization alternatively planned to carry out their working tasks in sudden case of IT distraction and loss of key operatives?

Stage – 3: reviewal of the network anatomy to make sure the organization is prepared against the external attacks (i.e. antivirus program, firewall) and evaluate the IT structure to spotify the maintenance of activities.

Stage – 4: it is ongoing stage and refers to the application and involves the follow-up assessments and process of implementing changes on application. Review the backups events, security systems being utilized in the organization such as antivirus. Reviews of the network facilities and related checks and policies. Review the authorization and rights according to needs. This stage could be happened annually with the agreement of the auditors but more often when the SMS is still maturing.

Stage – 5: in the very last of the audit, testing of controls might be occurred for the required control in the management system could be automated, semi-automated or manual.

10.16.1   BENEFITS OF ISO/IEC 27001

Organization's assets are of great importance. The standard ISO 27001 supports the establishments to manage the information security estate. However, the ISO27001 is the leading standard that fulfills the all requirements a security management system (SMS) needs [11] (Figure 10.12).

1. Enhancement of the business competence
2. Lessen the functional risks
3. Information security application insurance perceptively
4. Certification awarded to business partners and clients in order to get marketing initiatives
5. Most importantly aware the high-level managers and employees about security.

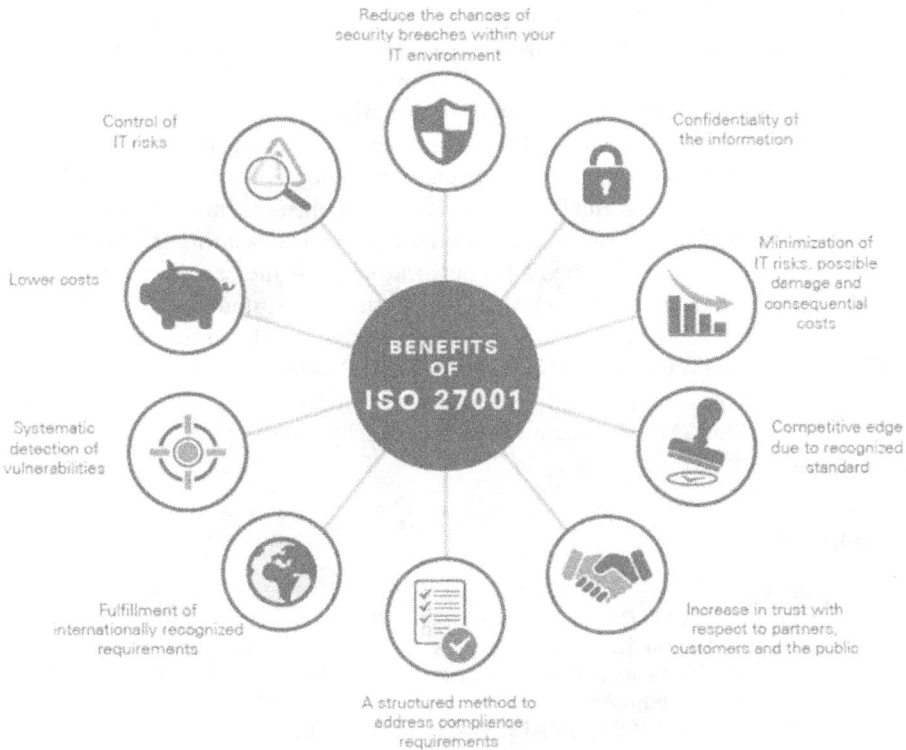

FIGURE 10.12   Benefits of ISO/IEC 27001 [38].

10.17   SECURITY MANAGEMENT SYSTEM CHALLENGES

There are numerous challenges related to risk assessment, which have been examined earlier. Now the challenges associated with the information security management system [39] will be explored next:

1. *Operational, procedure, and parameter challenges:* the contemporary information management system has played the major role in the running business environment. Information security breached down in the hard boundaries such as terrestrial, corporal, and analytical.
2. *The anthropoid Contest:* as the hackers employ time to identify the susceptibilities more than information security experts, so the humans are hard to administer in the framework of security of the organization.
3. *Fluctuating Executive Culture:* it is needed to have a somewhat better empathy of the communal characteristics of the enterprise security, exclusively the human factor. Also, the humans not like machines so the same information is not expected time by time.

10.18   CONCLUSION

In this perceptive environment and competencies, the enterprises must stay conscious about their assets, information security, and other relevant measures, otherwise they will fall behind in the race of technology and proficiencies. The feasibility and viability of the security management system (SMS) likely to be continuously grow as the obstruction in information technology systems tends to become critical and challenging. For the implementation of the complete set of rules and strategies, an organization should accommodate with the SMS with all of its formalities, such as quality assurance, risk management, risk assessment, and deployment of managerial-level security tactics. The security of information could be ensured in terms of the IT framework, which is the centered entity in the SMS. Security management in context of information systems needs to be prevalent so that the tactical issues be detected and avoided from the IT sector and allied with commercial governance mode. This chapter expounded the concept and working of SMS with the implementation of the ISO/IEC 27001 standard and acknowledged the SMS expansion to offer security and amplify corporate governance as well.

REFERENCES

[1] Alexander D., Taylor A., & Finch A. (2008). *Information Security Management Principles: An ISEB Certificate.* New York: BCS, the Chartered Institute.
[2] Jawad A., Hassan K. M., & Hussain F. (2015). Information security management for small and medium size enterprises. *Science International, 27*(3), 2393–2398. Retrieved from https://www.academia.edu/26971985/INFORMATION_SECURITY_ MANAGEMENT_FOR_SMALL_AND_MEDIUM_SIZE_ENTERPRISES on 10th November 2019.

[3] Buragga K. A., & Zaman N. eds. (2013). *Software Development Techniques for Constructive Information Systems Design.* IGI Global.

[4] Vacca J. R. (2010). *Managing Information Security.* New York: Syngress.

[5] Bourgeois D. (2019). *Information Systems for Business and Beyond.* Pressbooks.

[6] Retrieved from https://geek-university.com/ccna-security/confidentiality-integrity-and-availability-cia-triad/ on 11th November 2019.

[7] Ninja S. (2018). CIA triad. Retrieved from https://resources.infosecinstitute.com/cia-triad/#gref on 11th November 2019.

[8] Buckbee M. (2019). What is the CIA triad. Retrieved from https://www.varonis.com/blog/cia-triad/ on 11th October 2019.

[9] Qadir S., & Quadri S. M. K. (2016). Information availability: An insight into the most important attribute of information security. *Journal of Information Security*, 7, 185–194.

[10] National Academies Press. (1991). Concepts of information security. In *Computers at Risk Safe Computing in the Information Age.* N.W. Washington: National Academies Press. Retrieved from https://www.nap.edu/read/1581/chapter/4 on 4th November 2019.

[11] Abdaziz A. (2017). Information security management system. *International Journal of Computer Applications*, 158(7), 29–33. doi: 10.5120/ijca2017912851.

[12] Singh A., Vaish A., & Keserwani P. K. (2014). Information security: components and techniques. *International Journal of Advanced Research in Computer Science and Software Engineering*, 4(1).

[13] Kazemi U. (2018). A survey: Information security management system. *Journal of Analog and Digital Devices*, 2(3).

[14] Almrezeq N. (2021). An enhanced approach to improve the security and performance for deduplication. *Turkish Journal of Computer and Mathematics Education (TURCOMAT)*, 12(6), 2866–2882.

[15] Calder A., & Watkins S. (2019). The ISO 27001 risk assessment. In *Information Security Risk Management for ISO 27001/ISO 27002, third edition* (pp. 87–93). Ely, Cambridgeshire, United Kingdom: IT Governance Publishing. doi: 10.2307/j.ctvndv9kx.11.

[16] Carlson T. (2008). *Understanding Information Security Management Systems.* New York: Auerbach Publications.

[17] National Institute of Standards and Technology (NIST) Computer Security Division. (2010). A conceptual foundation for organizational information security awareness. *Information Management & Computer Security*, 8(1), 31–41.

[18] European Network and Information Security Agency (ENISA). (2010). ISMS framework. Retrieved from http://www.enisa.europa.eu/activities/risk-management/current-risk/risk-management-inventory/rm-isms/framework.

[19] European Network and Information Security Agency (ENISA). (2010). Risk management process. Retrieved from https://www.enisa.europa.eu/topics/threat-risk-management/risk-management/current-risk/risk-management-inventory/rm-process on 15th November 2019.

[20] Ali S., Hafeez Y., Humayun M., Mohd Jamail N. S., Aqib M., & Nawaz A. (2021). Enabling recommendation system architecture in virtualized environment for e-learning. *Egyptian Informatics Journal*, 2021.

[21] Carlson T., Tipton H. F., & Krause M. (2008). *Understanding Information Security Management Systems.* Boca Raton, FL:Auerbach Publications.

[22] Tashi, I., & Ghernouti-Hélie, S. (2009). Information security management is not only risk management. In 2009 Fourth International Conference on Internet Monitoring and Protection (pp. 116–123).

[23] Humayun M., & Jhanjhi N. Z. (2019). Exploring the relationship between GSD, knowledge management, trust and collaboration. *Journal of Engineering Science and Technology (JESTEC)*, 14(2), 820–843.

[24] Retrieved from https://www.comparethecloud.net/articles/what-is-iso-27001-and-why-do-i-need-it/ on 15th November 2019.

[25] Hamid M. A., Hafeez Y., Hamid B., Humayun M., & Jhanjhi N. Z. (2020). Towards an effective approach for architectural knowledge management considering global software development. *International Journal of Grid and Utility Computing*, 11(6), 780–791.

[26] Pavlov G., & Karakaneva J. (2011). Information security management system in organization. *Trakia Journal of Sciences*, 9(4), 20–25.

[27] TUV-NORD. (2015). Information security management system ISO 27001:2005. Retrieved from http://www.tuv-nord.com/cps/rde/xbcr/tng_in/Product_Information_27001.pdf on 2nd December 2016.

[28] BSI Group. (2017). BS EN ISO/IEC 27001:2017 – what has changed? Retrieved from http://www.bsigroup.com on 29th September 2019.

[29] TUV-NORD. (2015). Information security management system ISO 27001:2005. Retrieved from http://www.tuv-nord.com/cps/rde/xbcr/tng_in/Product_Information_27001.pdf on 23rd September 2019.

[30] Retrieved from https://en.wikipedia.org/wiki/ISO/IEC_27001#cite_note-4 on 4th October 2019.

[31] ISO/IEC 27001:2013. (2013). Information technology — Security techniques — Information security management systems — Requirements (second edition). Retrieved from https://www.iso27001security.com/html/27001.html on 18th November 2019.

[32] ISO/IEC 17799. (2005). _Information technology - Security techniques - Code of practice for information security management_.

[33] KARNBULSUK. (2009). Taking the first step with PDCA. Retrieved on 17th October 2019.

[34] Dionach. (2011). Update to ISO 27001 Planned for 2013. Retrieved on 20th October 2019.

[35] Retrieved from https://en.wikipedia.org/wiki/Security_controls on November 2019.

[36] Sinha M., & Gillies A. (2011). Improving the quality of information security management systems with iso27000. *The TQM Journal*, 23(4), 367–376.

[37] ISO. (2015). The ISO survey of management system standard certifications. Retrieved from http://www.iso.org/iso/the_iso_survey_of_management_system_standard_certi_cations_2015.pdf on 16th November 2019.

[38] TUV. ISO 27001 information security. Retrieved from https://www.tuv.com/philippines/en/iso-27001-certification.html on 21st November 2019.

[39] Ashenden D. (2008). Information security management: A human challenge? *Information Security Technical Report*, 13(4), 195–201.

11 Automatic Street Light Control Based on Pedestrian and Automobile Detection

R Sujatha, J Gitanjali, R. Pradeep Kumar,
Mustansar Ali, Ghazanfar, Baibhav Pathy,
and Jyotir Moy Chatterjee

CONTENTS

11.1   INTRODUCTION

An automatic street light control system is focused on energy saving and the use of street lamps efficiently when not needed. In our daily life, energy consumed by the conventional street lamps creates great energy demand. In the conventional street lights at night, the street lights will be ON with high intensity of light whether there is no vehicle movement or high vehicle traffic with low light intensity. To solve this issue, we come up with advanced techniques in street light control systems by implementing

DOI: 10.1201/9780367808228-11

our research work in automation of street lights based on vehicle detection by IR signals with the use of a microcontroller instead of using a microprocessor.

The lights detect the vehicle passing through roads to power on when the vehicle is almost sensed and switch off when no vehicle is detected. By doing this, power consumption is often saved, rather than lighting off when no vehicles are passing.

High-intensity discharge (HID) lamps are often utilized for streets. HID lamps are electrically gas-discharged but less energy efficient when compared to LEDs. IR LED transmit and receive signals to the 8051 microcontroller AT89S52 to perform switch ON and switch OFF or dim the intensity of light by 20% when no vehicles are detected. LED lights are being implemented in street arrangements where the IR sensors and photodiodes sense pedestrians or vehicles passing. The control logic is implemented in Microcontroller 8051 to regulate light-supported vehicle detection by two modes; bright and dim modes whenever there's any detection or no detection within the streets.

The project uses AT89S52 Microcontroller to read in the inputs from a series of IR transmitter and receiver signals. As you pass any object via the IR setup. Each IR setup will sense the object/vehicle and send a signal to the microcontroller. Based on vehicle presence, 8051 will turn on a set of white LEDs, which act like street lights. As the object passes by other IR setup, only a certain number of LEDs are turned on, keeping the energy consumption at a minimum.

Once the vehicle passes out of the last IR setup, all the LEDs turn OFF. This system is mainly concentrated in energy saving based on pedestrians or automobiles detected by the IR sensors with different phases of operation, namely dim, bright, and dark. By adopting this technology, we can overcome conventional street light systems.

The working of this proposed model is simple because when the pedestrian or vehicle passes the IR setup with blue & black LEDs passes information to AT89S52 microcontroller responses to the signal and passes instructions to power ON/OFF LED to glow or not.

11.2   BACKGROUND OF THE INVENTION

This invention is based on microcontroller and IR sensors for detecting pedestrians and automobiles. From this project, we can minimize the electricity wastage, we are using 8051 microcontroller, IR setup, and LEDs for implementation. The IR setup senses the vehicle pedestrian and sends received signals to the microcontroller 8051. Then that will help to control the LEDs lights whether to glow or not based on signals from the IR setup deciding whether to glow or not.

11.3   LITERATURE REVIEW

This system is based on the design of the modules, which help detecting the faulty street lamps of the system and give response to the street lights automatically [1]. In this automated street light system, it is purely based on the modules, which helps in controlling remotely the street lamps. This paper covers the study of the street lights by automated tracking, where the system can activate the increase in power management of the solar power generation system [2]. Output of this system is amplified based on the sun light density, which is being sensed by the sun tracking sensor sensing device. This system operates in an automatic manner that instantiates the street lights based on brightness and darkness [3]. The output of the system is surveyed based on vehicle movements and glow lights On/Off. This system works with the help of IR sensors along with Raspberry Pi [4]. The IR sensor will do the detection of vehicles or pedestrians, and then it will send responses to Raspberry Pi whether to switch On/Off street lamps. In this paper, LED lamps and Zig Bee modules were proposed by the author as a power-saving mechanism for street lights [5]. They have used LEDs rather than using the conventional street light for better high efficiency and less energy consumption.

In this paper, the author proposes an internet of things based system for traffic management since IOT is widely used nowadays [6]. IOT is the drive chain of future advancements in technologies. New methods of managing traffic based on the concepts of IOT. In this paper, the author presents a smart street light which makes use of infrared sensors technology [7]. At night or dark, the street lights automatically lighten up for a few seconds and then become dark. It adopts a dynamic control system, which gives response to dark and bright modes of operations. In this paper, the author proposes the Arduino technology to be used in street lighting for efficient functioning [8]. Modifications of the conventional street light system with deployment of Arduino technology makes this system more energy saving and efficient. The proposed system will be useful mainly in peak traffic and rainfall [9]. They use RTC where the street lamps will be glowing 100% intensity in high traffic time and lesser intensity when traffic is low. They have used light sensors to detect the traffic as well as weather. The main objective of the system is not lighting the street lamps when no objects are detected [10]. This system basically works by detecting the motion of the bikes, cars, trucks, and human beings, and it behaves according to the situation. At night when no objects are detected, the system goes to power off, and when objects are detected, it would perform the operation of glowing on the street lights.

This paper is based on an image analysis algorithm. In this, they have used two different techniques that are based on illumination of daytime and nighttime [11]. Images were taken, and choosing necessary algorithms, we can detect the vehicle at night as well as day. In this research paper, the detection of vehicles by an image taken by the camera is done [12]. In this system, the camera detects cars by finding the headlight and taillight by image segmentation and image pattern analysis method. This system is completely based on piezoelectric sensors to detect the movement of vehicles [13,14]. When a vehicle is detected by piezoelectric sensors, it instructs the street lights to glow On/Off. This system is completely dependent on piezoelectric sensors to perform operations. In this research paper, the system is based on digital image processing techniques [15]. In this, fast bright car images are segmented by automatic multi-level histogram thresholding approach to enhance image and analyze in night time. In this proposed system, the image sensor captures image frames and detects the headlight & tailight of the car or vehicle-based image segmentation approaches [16]. The image frames captured are detecting the vehicle by a spot that appears in the captured image.

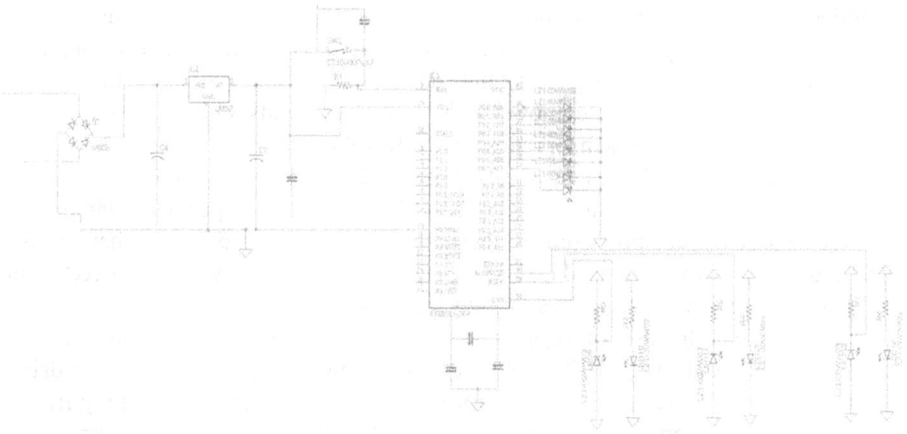

FIGURE 11.1 Circuit diagram.

In this paper, the system is widely used based on laser sensors and PIC microcontroller [17]. Laser sensors are implemented in the street, which detects the vehicle and gives signal to microcontroller to turn On/Off street lights. This research work is based on remote sensing of street light system. HID lamps are replaced by LED lamps in this system [18]. It is based on solar panels and an efficient microcontroller. The proposed system is based on low-cost microcontroller, rain sensors, laser sensors, and LEDs [19,20]. The management and controlling of the system is by microcontroller to take decisions based on lighting for the current situation. In this paper, a network-controlled charge transfer system is being implemented to transfer charge between the power grid and the vehicle that is mounted to the street light [21,22]. The local power grid and vehicle are connected by a wiring box, and when detected, it instructs to turn on street lights. From this research paper, it is based on a monocular precrash vehicle detection system [23,24]. It involves two algorithms to detect which are appearance-based hypothesis verification and multi-scale driven hypothesis generation, as shown in the Figure 11.1.

11.4   CIRCUIT DIAGRAM

11.4.1   Major Components Used

- MICROCONTROLLER 8051
- IR TRANSCEIVER AND RECEIVER LED Setup
- WHITE LEDS
- PCB BOARD COMPONENTS
- MICROCONTROLLER

The project uses AT89S52 microcontroller to read in the inputs from series of IR transmitter & receiver signals.

- IR SETUP:

As you pass any object via the IR setup, each IR setup will sense the object/vehicle and send a signal to the microcontroller.

- WHITE LED's:

Based on vehicle presence, 8051 will turn on a set of white LEDs, which act like street lights.

- As the object passes by another IR setup, only a certain number of LEDs are turned on, keeping the energy consumption at minimum.
- Once the vehicle passes out of the last IR setup, all the LEDs turn OFF.

COMPONENTS SOLDERED IN PCB BOARD:

- AT89S52-MICROCONTROLLER
- ON-OFF SWITCH WITH RED LED
- RESISTORS
- AC CABLE
- IR BLACK LEDs
- IR RECEIVER
- IR TRANSMITTER
- CAPACITOR.
- WHITE LED
- IR BLUE

11.5 ARCHITECTURE DIAGRAM

The path is the important thing, which we will be focused on in this system. Because creating the path (a small gap) in the printed circuit board between IR BLACK & BLUE LEDs will be the object detection done by the system. The WHITE LEDs will be in series connected with the IR transmitter and receiver, which will detect the object when it passes through the path and displays the output through white LEDs. A flowchart of the discussed proposal is shown in Figure 11.2.

11.6 PROPOSED WORK

In an automatic street light control system, the main components required to build the necessary circuit is Microcontroller AT89S52, IR setup for the white LEDs to glow brighter, dim or switched off. This system is completely focused on optimum utilization of street lamps with the main objective of saving energy. Because at night, most of the energy is wasted when there is no detection of pedestrians or automobiles, but the street lights still glow brighter. Keep this in mind as we build this system.

Building the circuit with necessary capacitors, resistors, power supply switch with red LED, AC cable, IR setup with blue & black LEDs that act as a path, white LEDs, and Microcontroller AT89S52 being soldered in PCB. The IR setup transmits and

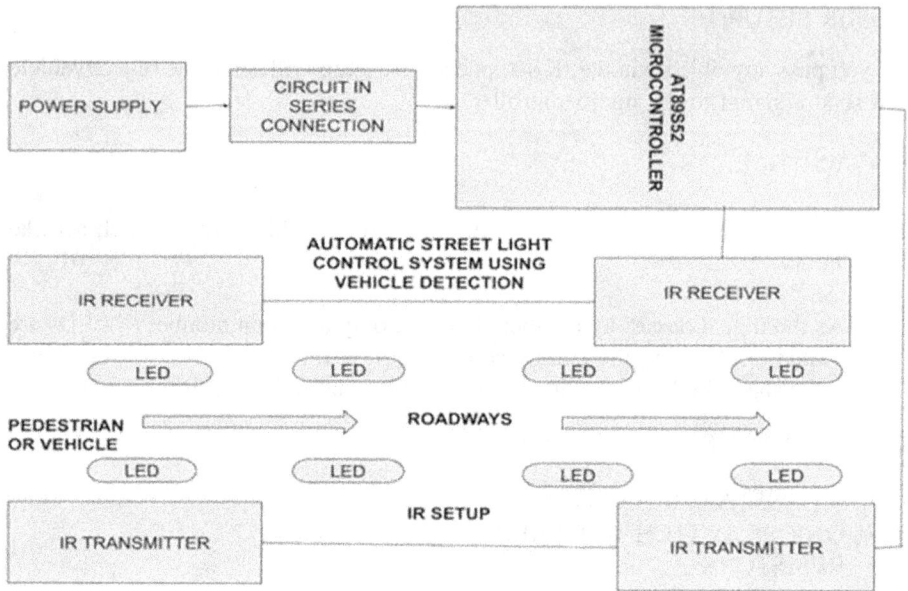

FIGURE 11.2 Flowchart.

receives signals from the path sending the information to the microcontroller. Based on the microcontroller instruction, the street lights perform modes of operation in projecting light in bright, dim, and dark. This is how our implemented system works by using IR sensors based on pedestrian or automobile detection on roads.

11.6.1 INSTRUCTIONS

- Connect the AC cable & turn the slide switch ON (the red LED should glow)
- Now pass your finger/any object through the passage between IR blue & black LEDs.
- As you keep passing your fingers, white LEDs will turn ON, illuminating the path ahead for you.
- Once you move out of the last IR setup, the top four white LEDs will be on for a few seconds and finally turn OFF.

11.7 RESULTS AND DISCUSSION

The proposed system uses AT89S52 Microcontroller with IR setup based vehicle detection. The images of the implemented system's IR setup and two phases of operation results are shown.

The circuit board have been built and necessary components have been soldered, as shown in Figure 11.3. After testing, we perform the operations in order to predict the output. In this system, it basically has two phases for street light system from dark mode to bright mode, as shown in Figure 11.4, as well as showing transition when the pedestrian moves, shown in Figure 11.5.

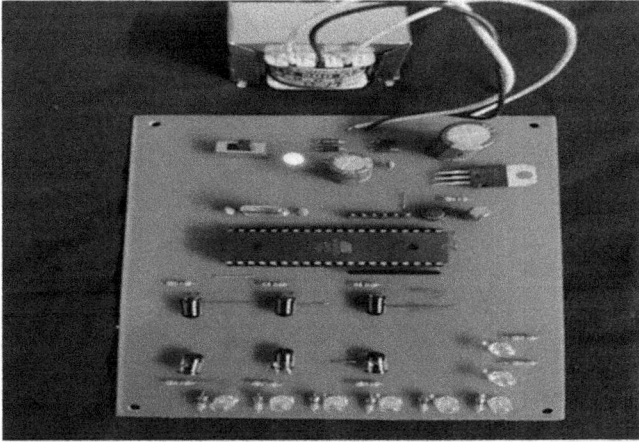

FIGURE 11.3 Automatic street light control system IR arrangement setup.

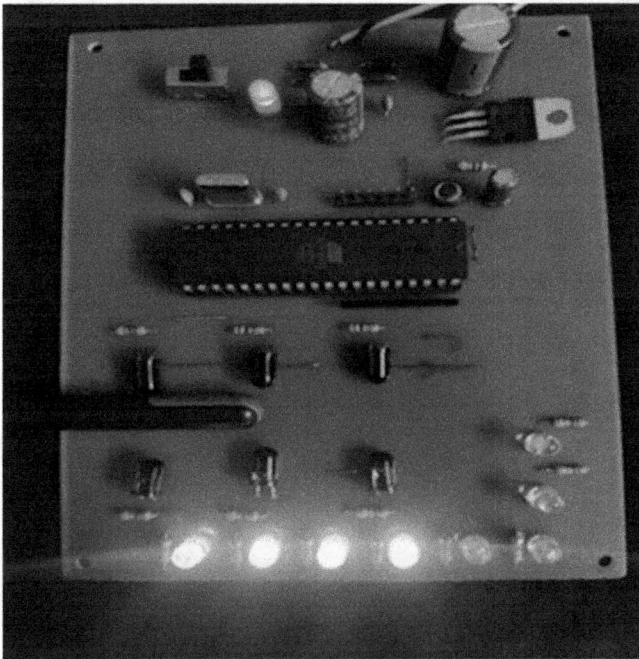

FIGURE 11.4 Phase I.

Street lights perform from dark mode to bright mode for fewer detection of pedestrian and automobile as shown in Figure 11.4. Street lights perform dull mode to bright mode for more detection of pedestrian and automobile in Figure 11.5. The automatic street light control system, when connected to the power supply, where 10 V DC voltage is given to the circuit where input is 5 V and the LEDs output is approximately 3 V, which is verified in voltmeter in Figure 11.6.

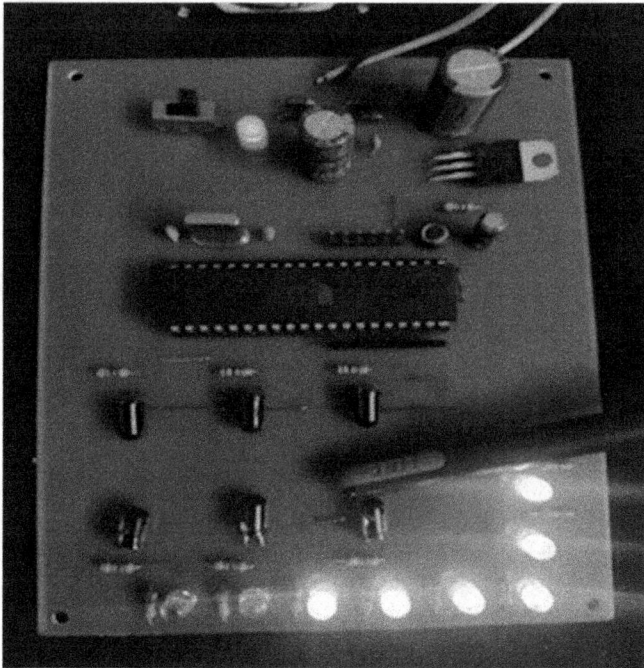

FIGURE 11.5 Phase II.

FIGURE 11.6 Measurement using voltmeter.

11.7.1   Usage of Automatic Street Lights

As the world is moving forward, the scarcity of energy is growing more and more. The hunger for energy is speeding up the animosity between different countries. In such a time, we need to prepare a variety of methods to conserve energy for us, as well as our next generation. Automatic street light is a method aimed to conserve a great amount of energy, which gets wasted even when there are not people in the street. It also helps the government to save a large amount of money, which gets used for providing street light to the general public, thus this money can be used to uplift poverty and other fields of public development and boost the country's economy.

The idea of automatic street light can be used for public safety by using them as traffic lights or as pole lights at night or during dark hours. These can help to avoid major accidents, as well as have labor-cutting costs, which are required in traffic posts.

11.7.2   Feasibility and Cost Effectiveness

The model proposed in this paper is very cost effective as compared to other models already present in the market. All the components listed in the model are easily and readily available in the local market and thus can be produced at mass scale without any further issue or complication. Thus, the feasibility of this model makes it more convenient for implementation of this model between the general public.

11.7.3   Used in Versatile Areas

This concept is being many more fields apart from street lights, such as in vehicles, automated doorways, public hotspots (schools and offices), etc. The adaption of this concept has a wide variety of usage in modern-day equipment and infinite possibilities in the future. For example, the switching of lights in classrooms when no one is there can save 762 MWh electrical energy annually, which is equivalent to 251 ton of less carbon released into the atmosphere, which can save $1.56 million that is wasted every year. So many researches have been going on where they use this idea to measure the amount of pedestrians on the road and by using complex algorithms and by using machine-learning approaches , the public crowding in the road is decreased and diverted. Safety-related projects have also been done to protect pedestrians, as well as avoid road accidents.

11.7.4   Large-Scale Manufacturing

As these things are made local, these can be built and assembled cheaply and efficiently. Production of these models should be boosted drastically so that they can be delivered to every local place in the country. This will increase and boost the country economy by showing country blooming with latest research and invention.

11.7.5 MERGING WITH AI AND IoT

Several projects have been ongoing to have a centralized traffic system, as well as predicting times of heavy crowds and the amount of energy required for every month. There has also been idea ongoing in installing renewable energy sources, such as solar panel and miniature windmills, which may power these street lights. Using these renewable energy sources with our designed model will make it free of cost within a few years (estimated seven years). Thus, this will be a major pro point of adopting this technology, as well as protecting the environment simultaneously. In artificial learning, work has been ongoing, such as self-learning and adaptivity, as well as reinforcement learning.

11.7.6 STABILITY AND EMERGENCY

Testing on stability of the product is further required. Backup plans should be made in advance to avoid any confusion or problem during the failure of the system. Emergency lights should be provided, which can be manually operated in case of sudden shutdown. Backup power systems like small investor should be provided to every energy grid in case of power system failure.

11.7.7 SECURITY PROTOCOLS

In this modern world, everyone is concerned about privacy and security. And so, we need to develop and adopt proper protocols for data transmission so that the traffic light and automated street lights don't get hacked by modern-day hackers and crackers. The concept of Zigbee protocol, which is highly used these days, should be adopted to transfer data packets from one lamp post to another. Also, we need to develop a protocol that is efficient enough to transfer enormous amounts of data to a long distance at a substantially low cost. For a remote place with lesser crowds, WiMAX (IEEE 802.16) can be deployed. All the channels should be secured and encrypted properly so that the hacker could not hack the micro controller. Also, the smart grid concept should also be used to save energy and security.

11.8 CONCLUSION & FUTURE WORKS

The working model we implemented is cost efficient, energy saving, and takes less installation time. As India wastes energy, up to 35–40% of energy through the legacy street light system has to be updated in terms of being an energy saving as well as a reliable source in case of maintenance and safety. This automatic street light control system based on vehicle or pedestrian detection is mainly focused on IR signals. The IR signals predict and perform operations like whether the lights have to glow brighter when they detect vehicles or pedestrians, dim when they detect vehicles or pedestrians, or off when not in use by effective AT89S52 Microcontroller with best IR setup. Each IR setup will sense the vehicle or pedestrian and send a signal to the AT89S52 Microcontroller. Based on the vehicle detected, 8051 will turn on a set of white LEDs, which act like street lights. As the vehicle or pedestrian passes by another IR setup, only a few LEDs are turned on, keeping the energy consumption at a minimum. Once the vehicle or pedestrian passes out of the last IR setup, all the LEDs turn OFF. In this paper, we have illustrated the street light control system in an effective as well as cheaper way by IR setup. In the future, when the IOT comes into play with more advanced IOT devices, then our implemented work will for sure help in taking steps forward toward the development of street light control systems.

Although the project has been completed, there is always a scope of improvement. This proposed model can be further improved by reprograming the micro controller and adding deep-learning agents and logistic flow algorithm. More work needs to be done to increase the efficiency of the sensor. More trails should be done with other varieties of micro controller. Simulation of different ways to arrange the PCB (printed circuit board) design, as well as making it as small as possible, should be tried out. Graphs should be made showing efficiency over the time.

REFERENCES

[1] Srikanth, M., & Sudhakar, K. N. (2014). Zigbee based remote control automatic street light system. *International Journal of Engineering Science*, 639–643.
[2] Bhuvaneswari, C., Rajeswari, R., & Kalaiarasan, C. (2013). Analysis of solar energy-based street light with auto tracking system. *International Journal of Advanced Research in Electrical, Electronics and Instrumentation Engineering*, 2(7), 3422–3428.
[3] Sheela, K. S., & Padmadevi, S. (2014). Survey on street lighting system based on vehicle movements. *International Journal of Innovative Research in Science, Engineering and Technology*, 3(2), 9220–9225.
[4] Sakthi Priya, V., Vijayan, M. M., & Sakthi Priya, V. (2017). Automatic street light control system using wsn based on vehicle movement and atmospheric condition. *International Journal of Communication and Computer Technologies*, 5(1), 06–11.
[5] Abrol, P. (2013). Design of traffic flow-based street light control system. *International Journal of Computer Applications*, 10911221.
[6] Saifuzzaman, M., Moon, N. N., & Nur, F. N. (2017, December). IoT based street lighting and traffic management system. In *2017 IEEE Region 10 Humanitarian Technology Conference (R10-HTC)* (pp. 121–124). IEEE.
[7] Sindhu, A. M., George, J., Roy, S., & Chandra, J. (2016). Smart Streetlight using IR sensors. *IOSR Journal of Mobile Computing & Application (IOSR-JMCA)*, 3(2), 39–44.
[8] Cynthia, P. C., Raj, V. A., & George, S. T. (2017). Automatic street light control based on vehicle detection using arduino for power saving applications. *International Journal of Electrical, Electronics and Computer Systems*, 6(9), 297–295.
[9] Khatavkar, N., Naik, A. A., & Kadam, B. (2017, August). Energy efficient street light controller for smart cities. In *2017 International Conference on Microelectronic Devices, Circuits and Systems (ICMDCS)* (pp. 1–6). IEEE.
[10] Beeraladinni, B., Pattebahadur, A., Mulay, S., & Vaishampayan, V. (2016, August). Effective street light automation by self-responsive cars for smart transportation. In *2016 International Conference on Computing Communication Control and automation (ICCUBEA)* (pp. 1–5). IEEE.

[11] Cucchiara, R., & Piccardi, M. (1999, June). Vehicle detection under day and night illumination. In *IIA/SOCO*.

[12] Chen, Y. L., Chen, Y. H., Chen, C. J., & Wu, B. F. (2006, August). Nighttime vehicle detection for driver assistance and autonomous vehicles. In 18th International Conference on Pattern Recognition (ICPR'06) (Vol. 1, pp. 687–690). IEEE.

[13] Abinaya, R., Varsha, V., & Hariharan, K. (2017, February). An intelligent street light system based on piezoelectric sensor networks. In 2017 4th International Conference on Electronics and Communication Systems (ICECS) (pp. 138–142). IEEE.

[14] Humayun, M., Jhanjhi, N. Z., Alamri, M. Z., & Khan, A. (2020). Smart cities and digital governance. In *Employing Recent Technologies for Improved Digital Governance* (pp. 87–106). IGI Global.

[15] Chen, Y. L., Wu, B. F., Huang, H. Y., & Fan, C. J. (2010). A real-time vision system for nighttime vehicle detection and traffic surveillance. *IEEE Transactions on Industrial Electronics*, 58(5), 2030–2044.

[16] López, A., Hilgenstock, J., Busse, A., Baldrich, R., Lumbreras, F., & Serrat, J. (2008, October). Nighttime vehicle detection for intelligent headlight control. In International Conference on Advanced Concepts for Intelligent Vision Systems (pp. 113–124). Springer, Berlin, Heidelberg.

[17] Singh, M. A. L. G. M., & Husain, M. D. F. Review on vehicle movement based street lights.IJSRD - International Journal for Scientific Research & Development| Vol. 4, Issue 11, 2017, pp. 365–366.

[18] Sarma, A., Verma, G., Banarwal, S., & Verma, H. (2016, March). Street light power reduction system using microcontroller and solar panel. In 2016 3rd International Conference on Computing for Sustainable Global Development (INDIACom) (pp. 2008–2010). IEEE.

[19] Husin, R., Al Junid, S. A. M., Abd Majid, Z., Othman, Z., Shariff, K. K. M., Hashim, H., & Saari, M. F. (2012). Automatic street lighting system for energy efficiency based on low-cost microcontroller. *International Journal of Simulation Systems, Science & Technology*, 13(1), 29–34.

[20] Garg, S., Chatterjee, J. M., & KumarAgrawal, R. (2018, August). Design of a simple gas knob: An application of IoT. In 2018 International Conference on Research in Intelligent and Computing in Engineering (RICE) (pp. 1–3). IEEE.

[21] Lowenthal, R., Baxter, D., Bhade, H., Mandal, P., & Tormey, M. T. (2011). *U.S. Patent No. 7,952,319*. U.S. Patent and Trademark Office, Washington, DC.

[22] Hamid, B., Jhanjhi, N. Z., Humayun, M., Khan, A., & Alsayat, A. (2019). Cyber security issues and challenges for smart cities: A survey. In 2019 13th International Conference on Mathematics, Actuarial Science, Computer Science and Statistics (MACS) (pp. 1–7). IEEE.

[23] Sun, Z., Miller, R., Bebis, G., & DiMeo, D. (2002, December). A real-time precrash vehicle detection system. In Sixth IEEE Workshop on Applications of Computer Vision, 2002 (WACV 2002) (pp. 171–176). IEEE.

[24] Humayun, M., Jhanjhi, N. Z., and Alamri, M. Z. Smart secure and energy efficient scheme for E-Health applications using IoT: A review. *International Journal of Computer Science and Network Security,* 20(4), 55–74.

12 Cost-Oriented Electronic Voting System Using Hashing Function with Digital Persona

Muhammad Talha Saleem, Noor ul-Ain, and Zartaj Tahir

CONTENTS

12.1   INTRODUCTION

"Democracy means that Administration of the individuals, by the individuals, for the individuals" according to Abraham Lincoln. A majority-rules system is the arrangement of government by the entire populace or all the qualified individuals from the state commonly through chosen agents [1]. The concept of democracy is the simple and straightforward development of individuals. Now, development has its own classifications and modes to achieve. Individuals can also be classified as minorities, native, immigrant, etc. Each individual has different financial conditions, strict and social convictions, and land blocks. Majority-rules systems remember these and give an *equity* in chances to the denied class. Dr. Ambedkar's view further explains the concept of democracy. He says that democracy is a manifestation of the development of citizens having equality, liberty, and fraternity in the core of the *means* to achieve an *end* development [2]. The main role of democratic government is to decree that superiors and inferiors are the same. In democratic government, the public all have the right to give their opinion using a systematic process.

The defining feature of a democracy is the participation of its citizens in the electoral process [3]. The public elects a person for a particular seat, and if the pursuance of the political leader is not satisfactory, the public has the right to change that leader. The more citizens who vote, the more the democracy is expected to be representative [3].

Elections play a substantial role in democratic government. The absence of free and fair elections in government can prompt the destabilization of a nation [4]. Over the last few years, Pakistan is endeavoring to conduct effective, fair, free, and transparent elections that may result in selecting a competent leader who will lead the public desire accordingly. A responsive government that does not intefere plays an important role in providing a satisfactory election environment. Great governance in Pakistan relies upon free and straightforward decisions. The fundamental aspect of a vote-based framework and great administration are the electoral races themselves. It is not feasible to consider a vote-based political system without direct and straightforward, free and reasonable decisions in light of the fact that a people's intensity can't develop without giving them a reasonable chance to pick their delegates [4].

Like other developing countries, it's also the responsibility of our government to resolve the systematic problems that occur in the phase of elections and diminish our turn out. Pakistan challenged numerous matters in election procedures [4]. Rigging is the main problem of elections, and the results negatively impact the economy and repute of the country, with Pakistan facing many challenges and problems in front of other countries. Electoral rigging has hindered Pakistan's democratic development, worn political stability, and contributed to the collapse of the rule of law [5].

Numerous factors support the rigging practices, but the substantial one is the traditional paper-ballot voting system. It allows access of any unauthorized person to temper the election results. Election officials are also expected to notice and report fraud [6]. When the democratic election process is finished, and all of the voters have left, how simple is it to get to the polling forms? The issues with paper voting forms are the dangers of over casting a ballot, under casting a ballot, hanging chads, and so on [7].

Functional political arrangements for free, fair, and crystal-clear elections are a significant ingredient [4]. To resolve all these issues, it is important to establish a reliable voting system that overcomes all the fraudulent practices during the election process. Our present voting system registers voters by fingerprint authentication before the elections. At the time of the election, voters cast their vote by fingerprint verification. This fingerprint authentication ensures that only authorized people can cast the vote. Voters can cast votes at once. Voters can select the unsatisfactory option (vote no) during vote casting if they are not satisfied with any listed candidate. If the majority of voters selects the not satisfied option in the election phase, elections will be re-conducted on that particular area. During the election, votes will be counted by an autocounter process. This will reduce the result compilation time. After voting sessions, results will be displayed numerically as well as graphically (pie chart, bar chart).

12.2 LITERATURE REVIEW

This paper is presented by the writer (Olaniyi, O.M., *et al.,*) V-Authenticate: Voice System Authentication System for Electorates Living with Disability 2019). Some 0.15% of the total populace lives with incapacity; that is about one out of each seven people. There is no appropriate mechanism for large-volume electorates living with incapacity to completely partake in the appointive procedure; this lack deflects popular government from giving this subset of people a decision in how they wish to be represented. Along these lines, we present V-authentication, a voice acknowledgment and confirmation framework to address the issue of verification for the impaired electorate by accepting their voice as a quantifiable biometric quality. The voice acknowledgment framework deals with the confirmation and checks votes of PWD voters (people with disabilities). The sign investigation of the voice signal was cultivated by utilizing MFCC strategy on voice range factors [8].

This paper is presented by the writer (Nakirya Brenda Kintu, M.I.Z., a secure e-voting system International Conference Proceeding-International Systems and Engineering, 2018). Voting tactic is not secure owing to the attributed restrictions such as repeated voting and vote miscounts during the post-election phase of the voting process. The proposed framework has an auto count and grading highlight that naturally includes the votes; when a vote is thrown from any surveying station, afterward, the framework reviews the applicants as needed. The framework settles the security issues identified regarding genuineness, uprightness, and respectability during decisions [9].

This paper is presented by the writer (Shaheen, S., Yousaf, M., & Jalil, M., A Smart Card Oriented Secure Electronic Voting Machine Built on NTRU. 2018). Proposed EVMs (electronic voting machines) don't address all the security issues and prerequisites. They don't display an expansive perspective on secure EVM, which can address almost all security necessities. So the proposed system expands on another safe and productive EVM dependent on the lopsided cryptosystem, called NTRU. System-utilized smart cards and biohash fingerprints are used for voter verification, and these safeguard the voter's secrecy by utilizing blind signatures, just as votes are checked through the homomorphism count process. The proposed applied electronic voting scheme is secure, straightforward, and proficient for huge-scale decisions [10].

This paper is written by the writer (Sreerag M, S.R., Vishnu C Babu, Sonia Mathew, Reni K Cherian, Aadhaar Based Voting System Using Android Application. International Research Journal of Engineering and Technology (IRJET), 2018). Illicit democratic process is the principle issue looked by the current system. Different issues like the cost for directing elections, sitting tight for results affirmation, an excess of paperwork, and the possibility of human blunders are likewise the primary hazardous components for leading transparent elections. To defeat these problems, the authors propose an Aadhaar based democratic framework, which is an Android-based application that gives the democratic framework two key highlights; expenses are decreased in the elections and counterfeit votes are avoided. The uniqueness of this system is the utilization of biometrics, which aids in deciding if the individual voter is legitimate or not [11].

This paper is written by the writer (Hjálmarsson, F.Þ., et al. e-voting Blockchain-based system. IEEE 11th International Conference on Cloud Computing (CLOUD) 2018). Blockchain base e-voting systems is dependent on public blockchain, which is ineffective concerning money-related expenses. Another hazard is the high traffic in the network, which could influence the throughput of votes in the system, making it less time-efficient. 51% of attack poses a threat to a public blockchain. In this work-in-progress paper, we assess the use of square chain as support to execute dispersed electronic systems that use shrewd agreements to empower secure and cost-productive political elections while ensuring voter privacy. Utilizing an Ethereum private blockchain, it is conceivable to send transactions every second onto the blockchain, using each part of the brilliant agreement to facilitate the heap on the blockchain [12].

This paper is written by the writer (Najam, S.S., A.Z. Shaikh, and S. Naqvi, A novel hybrid biometric electronic voting system: Integrating finger print and face recognition. Mehran University Research Journal of Engineering & Technology, 2018). This recently proposed system utilized face recognition and fingerprint techniques to secure the e-voting system. But the facial recognition system used and employed is quite in-effective having a success percentage of only 58% and a response time of 15 seconds. Besides, it lacks data encryption or security for the secrecy of the ballot. To give better outcomes in contrast with single-identification-based systems, we create and present an electronic system to annihilate deceitful practices by including twofold client-recognizable proof checks; for example, this process uses facial recognition and unique marks. Facial recognition is cultivated through an element extraction-based AI calculation, while unique mark-based recognizable proof is accomplished through the example recognition technique [13].

This paper is written by the writer (N.AdityaSundar, P. M.V.Kishore, and Ch.Suresh, A Secure E-Voting System Using RSA and Md5 Algorithms Using Random Number Generators *International Journal of Applied Engineering Research* ISSN 0973-4562, 2018). Previously proposed e-voting system based on public key encryption algorithm RSA which contains three parts: login, voting and election administrator server. But the system has disadvantages like there is no any e-registration and more expensive cost and communication overhead due to RSA algorithm. In this paper, for the implementation of secure e-voting, RSA and MD5 algorithms have been used, which use public key encryption cryptographies and hash functions via digital signatures to secure and efficient casting of vote. RSA is used for the e-registration of the clients. MD5 calculation is used in the confirmation and vote-transfer stage [14].

This paper is written by the writer (Darwish, A. and M. Gendy, A New Cryptographic Voting Verifiable Scheme for E-Voting System Based on Bit Commitment and Blind Signature. *Int. J. Swarm Intel. Evol. Comput*, 2017). Numerous nations have not executed internet voting through I-casting a ballot systems. I-voting is a reprimanded and profoundly intricate process. Voting through the internet is an attractive process, yet with the utilization of the internet, the process will face new dangers. There are a few assaults that should be considered before adopting internet casting ballot plans. The paper introduces another cryptographic voting system that depends on the public key infrastructure (PKI), blind digital signature dependent on RSA cryptosystem and Bit commitment plans (BCS) as a part of cryptographic conventions. The proposed system secures against outstanding assaults, blunders, and electronic misrepresentation [15].

This paper is written by the writer (Rahul Dongre, P., Shubham Pal, Sangeeta Dhurve, Naina Surywansi, KrunalItankar A Review Paper on Fingerprint Based Voting Machine. *International Journal of Advanced Research in Electrical, Electronics and Instrumentation Engineering*, 2017). To make the quick, secure and effective democratic procedure utilizing topographical UI. The primary goal of this paper is structure and advancement of a Fingerprint Electronic Voting System to check qualification by Comparing current unique mark with the one previously put away in the system's database, and afterward they will be permitted to make their choice utilizing cordial geological UIs, secure and effective democratic procedure utilizing land UI [16].

In this paper the writer (Balkrushna Bhagwatrao Kharmate, S.S.S., Prashant Ravindra Kangane,Tushar Anant A Survey on Smart E-Voting System Based On Fingerprint Recognition. International Journal of Innovative Research in Computer and Communication Engineering, 2015). In voting systems, the security of the framework depends for the most part on the polling stations. Security of information, protection of the voters, and the precision of the vote consider primary perspectives that must be considered. Separation is an additionally tricky figure present in current voting systems. To defeat these issues, we create a web-based voting system utilizing fingerprint recognition. The voter authentication in online procedure should be possible by UIDAI information network and enrollment by unique finger-impression acknowledgment. Voters make their choice from any voting booth or spot for their registered voting area. This methodology is expanding the democratic rate. Also, keen security helps to stop phony or compromised democratic processes [17].

This paper is written by the writer (Tornos, J.L., J.L. Salazar, and J.J. Piles. An e-voting platform for QoE evaluation. In 2013 IFIP/IEEE International Symposium on Integrated Network Management (IM 2013). The proposed voting system provides the anonymity which prevents making a link between each voter and ballot. But when this feature is needed it is necessary to find a way to link together the ballots of a voter without revealing the actual identity of the voter. We built a safe e-voting system in view of ring signatures providing numerous features, for example, linkability or anonymity. The framework likewise permits connecting together every one of the voting forms of every client, without loss of secrecy, and viewing the various patterns in the clients' assessments, in this way helping the data-gathering process in QoE assessment [18].

12.3 PROPOSED METHOD

The intent of the system prototype is to build a model of a proposed system by the hardware and software integration, which performs all the functionalities efficiently and accurately according to the system requirements (Figure 12.1).

12.3.1 System Requirements

Our e-voting system includes functional as well as security requirements. The system supports the verifiability, flexibility, transparency, eligibility, security and privacy factors, but here we limit our discussion only into five main requirements.

Authentication: Only the validated or certify voter can cast their vote.

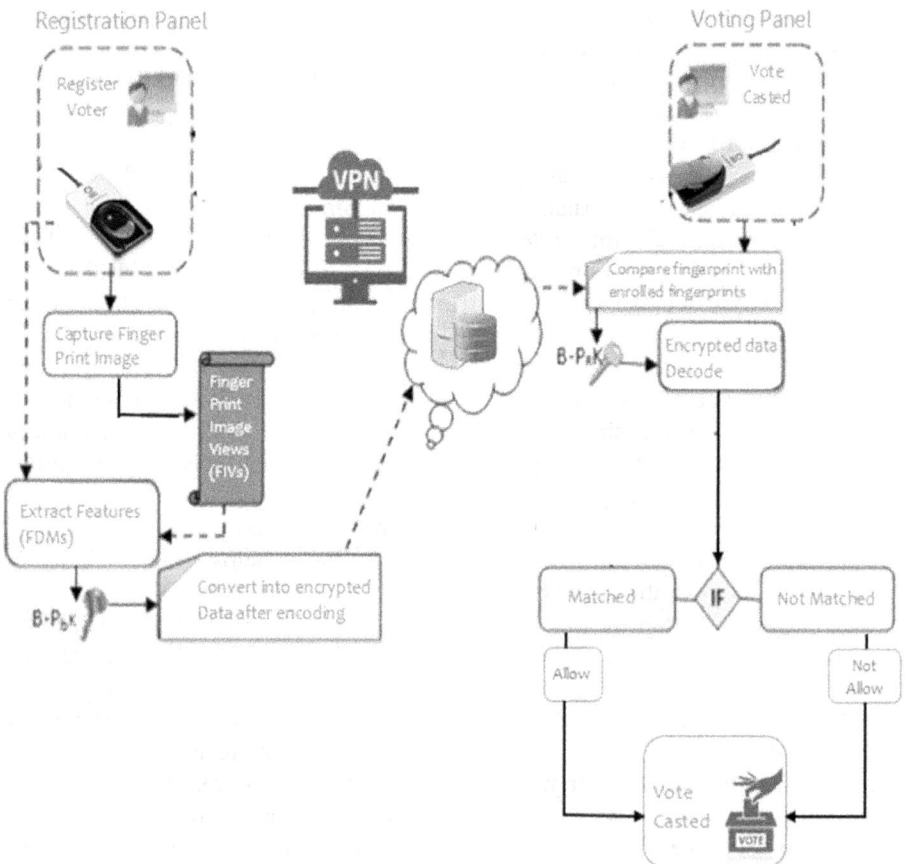

FIGURE 12.1   Election system flow.

Integrity: Nobody is able to modulate the data once it is recoded. System also ensures that the data cannot be altered by any user.

Confidentiality: Only an authentic person can access the structure, and the anonymity of the voter remains preserved.

Accuracy: System ensures the least error processing in counting the return accurately, and the generated results are unambiguous.

Uniqueness: The voter will be able to cast their vote only once.

12.3.2 SYSTEM ARCHITECTURE

The proposed electronic democratic framework was created to enable the overall population to cast a ballot through a work station; the framework is an open-finished sort that obliges both the director and the voter [19]. Here below presents the architecture of system (Figure 12.2).

In our proposed e-voting system, digital data of voter and candidate confidentially are transferred. All the data during registration and voting is saved in a database, which is centralized; the data SQL database is used for storing the data. Voters reach a centralized election platform and register their vote using fingerprint impressions via U.are.U digital persona biometric device. Voters also receive confirmation messages through SMS gateway services.

12.3.3 SYSTEM MODULES

The system model develops by the integration of hardware and software to acquire the better results for the proposed e-voting system.

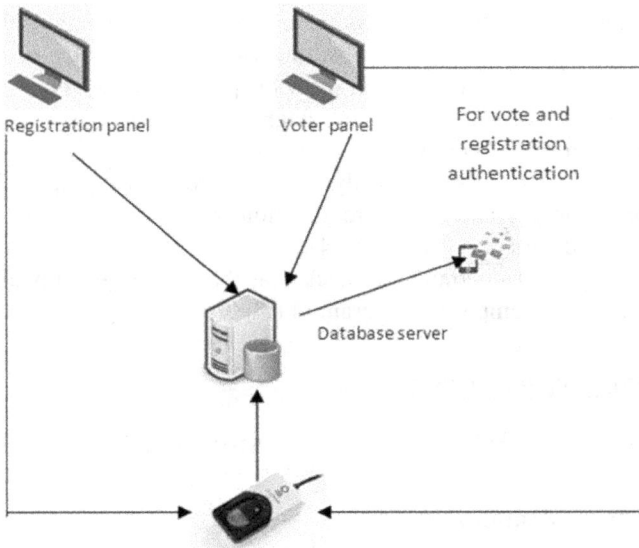

FIGURE 12.2   Architecture of e-voting system.

Authentication Module: Every human being has unique fingerprint impressions. Therefore, fingerprint impressions can most likely be used securely for the purpose of person authentication. Fingerprint identification is based primarily on the minutiaes , or the locationand directionof the ridges endings and bifurcation (splits) along ride path [20]. The verification of the fingerprint is done by the fingerprint module. Fingerprint module captures the image of minutes and converts into binary data. These binary data are stored into the database and after comparing, it verifies the person's identity. The proposed e-voting system integrates U.are.U digital persona fingerprint device to the software of the system.

Functional Modules: Functional modules are specified according to the end user. There are three functional modules: administrator module, voter module, and server module. Here below we discuss it briefly.

Administration module: Refers to the admin officer who controls the overall election system. The administration module manages the registrations of the voter and candidate, manages the concluding results, generates overall reports, and views the detail of casting votes.

Voter module: Refers as voter panel where voter cast their votes. It supports the entry of the voter's vote number. Voter verifies their authenticity via thumbprint impressions. After the verification process, voter elects a candidate according to their choice.

Server module: Connected both with administrator and voter modules. It acts like a centralized database where all the data are stored. All modules are associated with this centralized database system.

12.3.4 SYSTEM MODELING

System modeling demonstrates the route toward making dynamic models of a system, with each model showing a substitute view or perspective of that system. We use different UML diagrams, which show the pictorial view of our system, such as: ER diagram, activity, and component diagram.

The ER diagram of e-voting system is (Figure 12.3).

The activity diagram presents the dynamic nature of our proposed system. It represents the whole procedure from registration of voter through the submission of their vote. The diagram is (Figure 12.4).

We draw component diagrams to check that that all the functionalities of our system are done. The component diagram is (Figure 12.5).

12.4 WORKING PROCEDURE

Here we discuss the working procedure of the e-voting system.

12.4.1 ELECTION PARTICIPANTS

The proposed system includes two main participants for the creation of an election process.

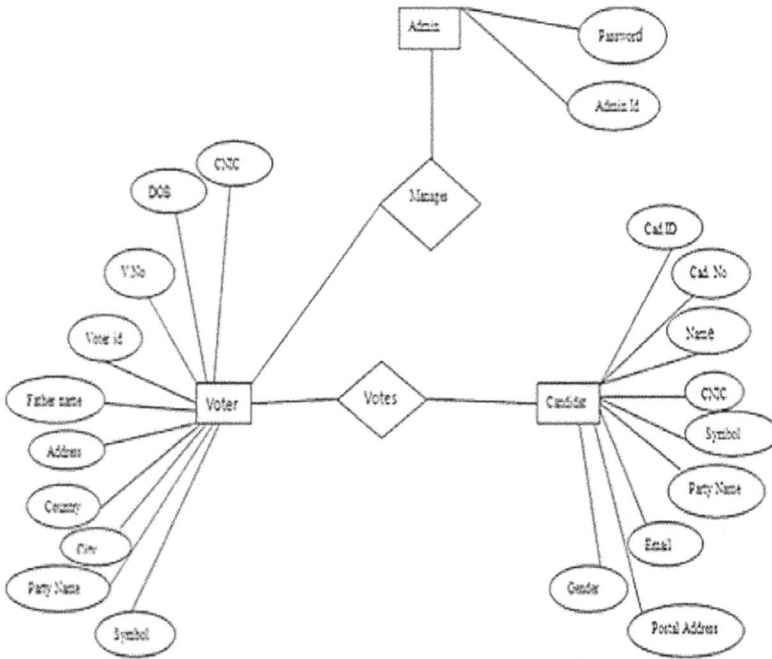

FIGURE 12.3 ER diagram of e-voting system.

Election Administrator: The election administrator is the admin office that controls the overall election process and activates or deactivates the election session. It also manages the results and registration process of the voters and candidates.

Voter: The voter is that eligible participant who is able to cast the vote at election time by authentication of their fingerprints and selects the candidate according to their choice.

12.4.2 ELECTION PROCESS

Election process comprises four main steps: registration process of the voter, activation of election, vote casting, and announcement the results.

Voter Registration: The registration of the voter starts on a specific day. Voter arrives at a centralized registration platform to register him/herself using a biometric device. The voter places a finger on a device that captures fingerprint minutes and stores them in a database. A vote number is given to voter and conformation message is send for surety of voter (Figure 12.6).

Election Activation: Election activation is a process in which the administrator declares the candidate's designation and sets the elections period by providing election date with starting and ending time (Figure 12.7).

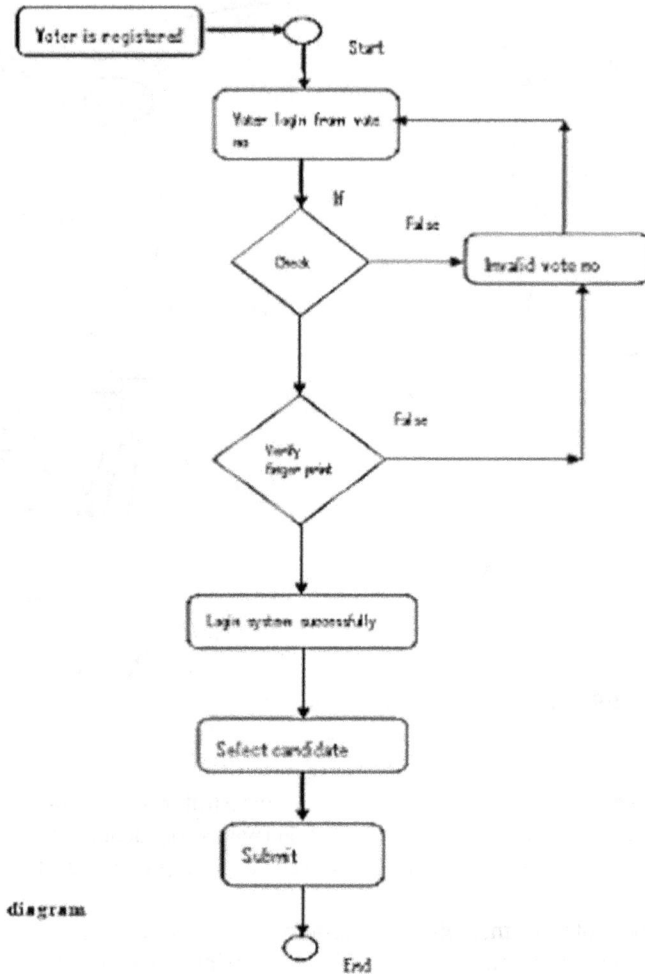

FIGURE 12.4 Activity diagram of e-voting system.

Voting: At election day, the voter cast votes to their desired candidates. Voter enters their assigned vote number (Figure 12.8).

System displayed voter's information and request for their thumb prints (Figure 12.9).

A fingerprint scanner scans thumbprints and tries to find the matching templates into the database (Figures 12.10 and 12.11).

After thumbprint verification, the candidate selection page is displayed where the voter selects their desired candidate and submits their vote. (Figure 12.12).

If voter cannot cast their vote, this active session will expire, and the voter will again need to login into the system. After the vote submission, a

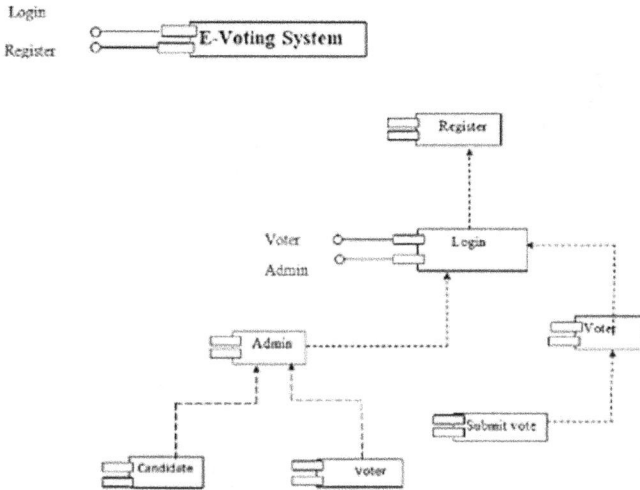

FIGURE 12.5   Component diagram of e-voting.

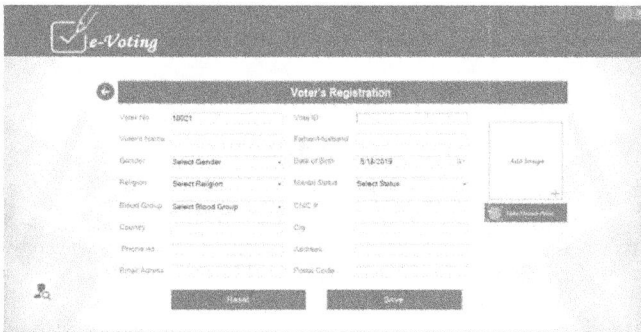

FIGURE 12.6   Voter registration panel.

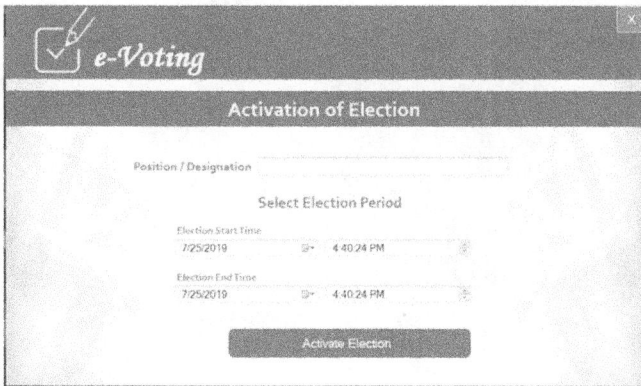

FIGURE 12.7   Activation of election.

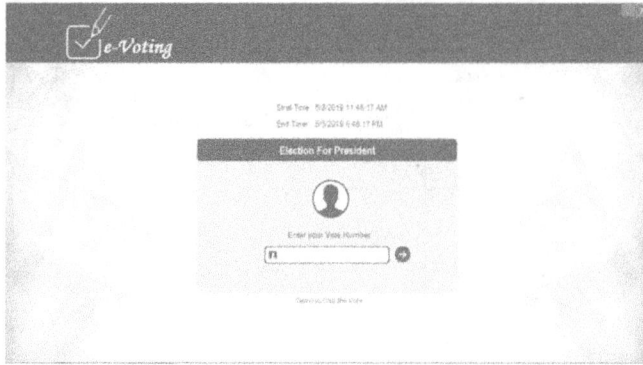

FIGURE 12.8 Voter panel.

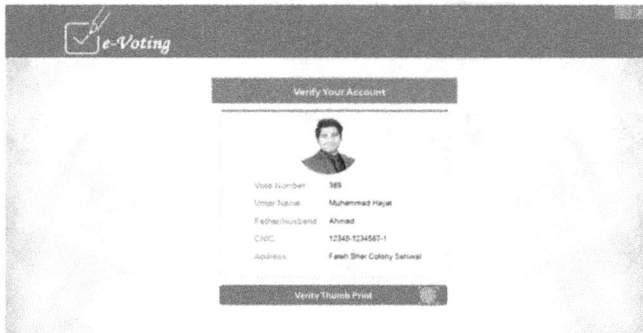

FIGURE 12.9 Voter profile.

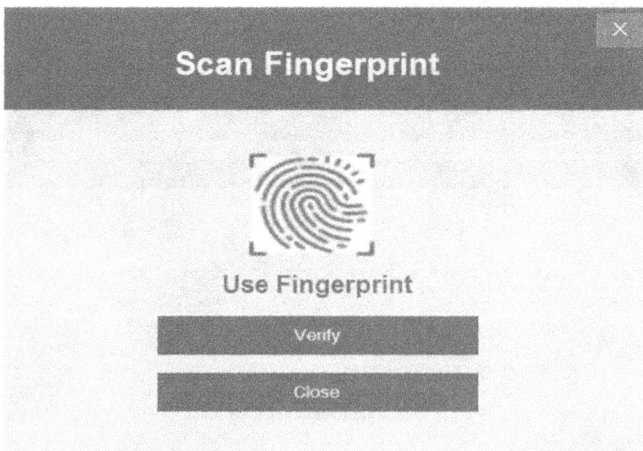

FIGURE 12.10 Scan fingerprint.

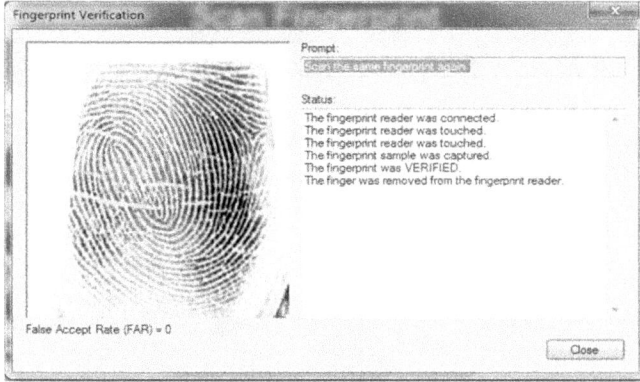

FIGURE 12.11 Fingerprint verification.

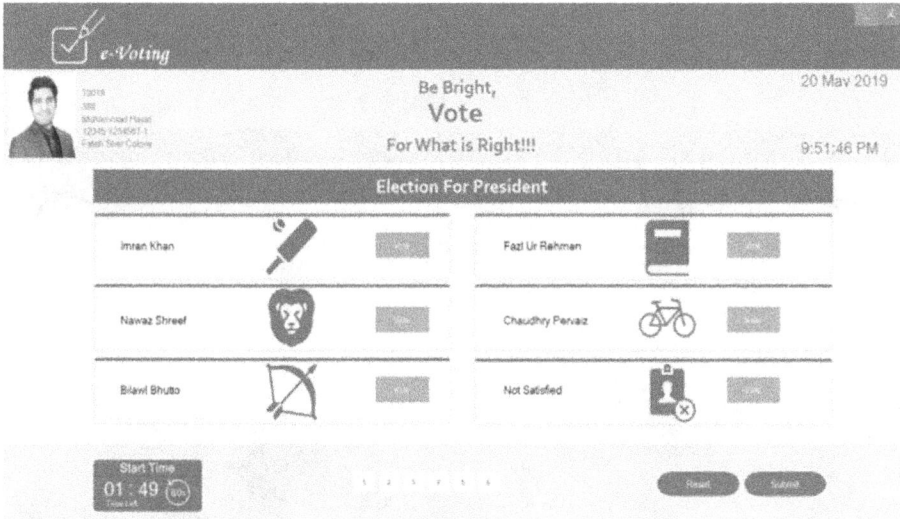

FIGURE 12.12 Candidate selection page.

verification message will be sent to the voter confirming that their vote has been cast.

12.4.3 GENERATE RESULTS

System generates effectual results through an autocount process. An autocount process automatically computes a cast vote and displays results in pictorial, graphical, and tabular form (Figures 12.13–12.15).

System also generates overall election reports of candidate and voters (Figures 12.16 and 12.17).

FIGURE 12.13 Results page.

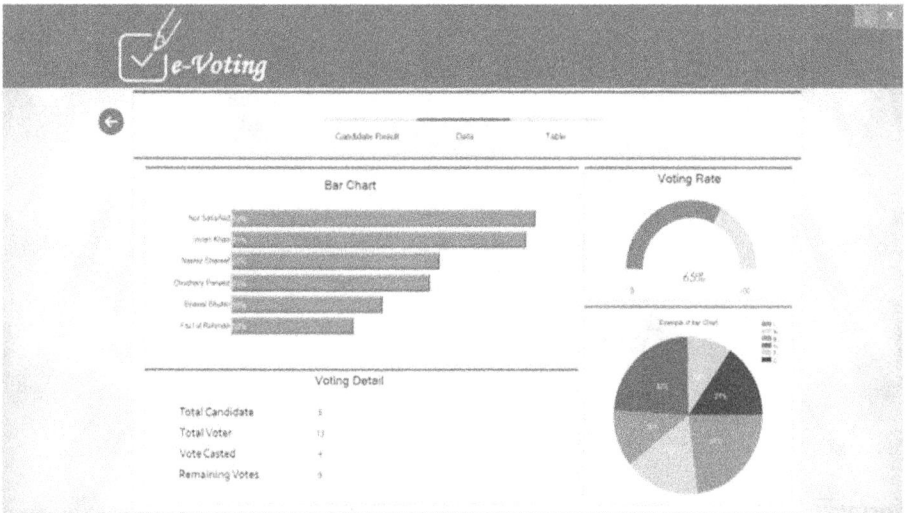

FIGURE 12.14 Graphical view of result.

12.5 DISCUSSION AND FINDINGS

12.5.1 OUTCOME OF PROPOSED VOTING SYSTEM

This system provides the cost-oriented, highly secure, less human effort oriented process that provides reliable results and avoids the biasness for the user (approach, database administrator); this real-world digital solution

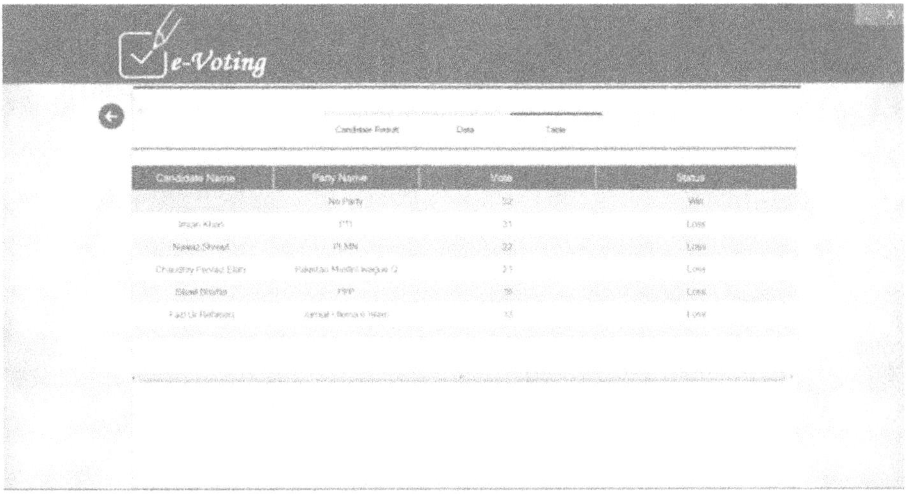

FIGURE 12.15   Tabular view of result.

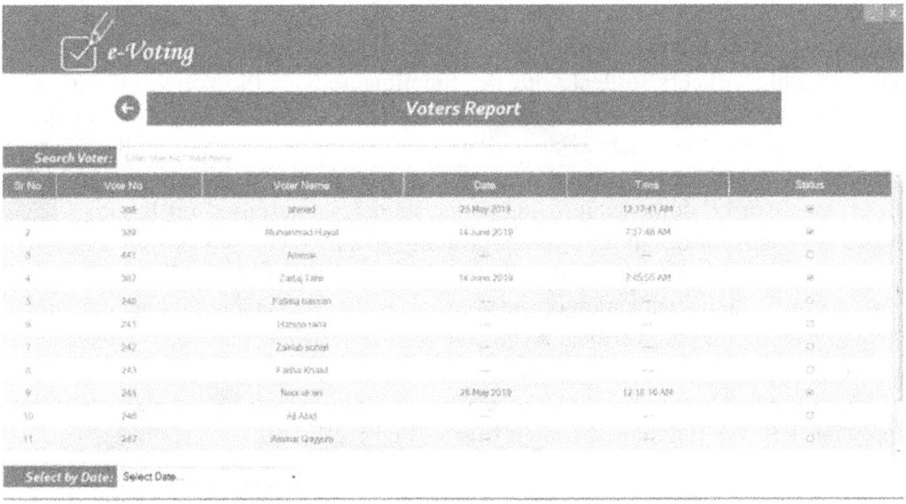

FIGURE 12.16   Voter report.

improves the selection of candidates by reducing the raging contradictions over the opponents. The digital records can be stored more efficiently than a physical record. Communication via virtual private network (VPN) facilitates the system to fulfill the confidentiality and availability of the resources nearest to the user. The opinion of the voter integrated at the user end will transfer in an optimized way, smoothing computation and leading to the success of the system.

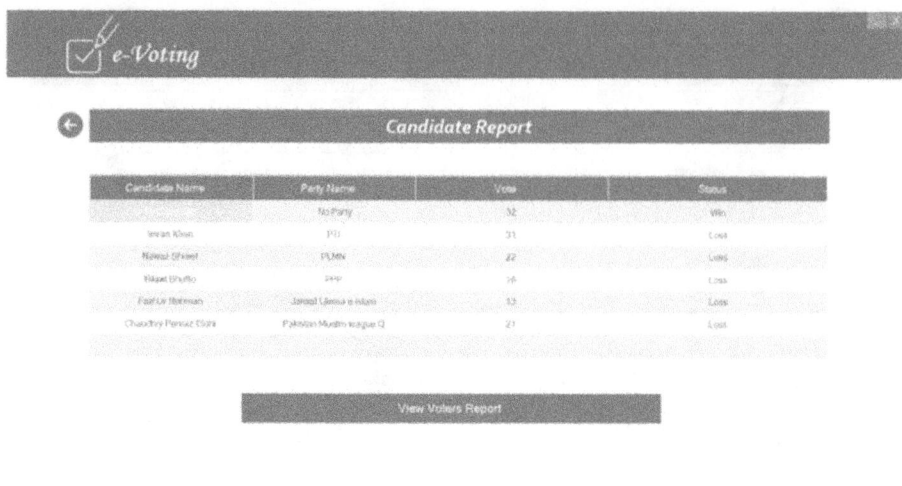

FIGURE 12.17 Candidate report.

12.5.2 EXISTING VOTING SYSTEM OF PAKISTAN WITH PROPOSED SYSTEM

The existing voting system works manually using ballot paper and ballot box, along with presiding officers to deal with the functionalities of the voting system. The proposed system is user-oriented, raging-free with less effort and optimized cost. It omits a platform that deals with digital data rooting to the domain base server environment using virtual private network. No need for physical ballot box or ballot paper; this can be done while using the system (system based on hashing using digital persona). In our system, fingerprinting of the vote caster is stored in a database in encrypted form using hash function; at the other end, the decryption takes place for providing verification of the vote caster. For enhancing fake modification, the user identity mobile verification feature is also implemented; vote casted can be enabled once but cannot be updated.

The stored pictorial view of the encrypted data shown in the database after applying the hash function is shown below (Figure 12.18).

12.5.3 OPTIMIZATION IN MULTIPLE ENHANCEMENT FACTORS IN PROPOSED SYSTEM

Enhancement in the existing system reaches the optimal result on the basis of facts. Our system contains the multiple factors that make the process more efficient and cost effective while enhancing the security factors.

12.6 CONCLUSION

This paper represents the secure and rigging-free electronic voting system by using the technology of U.are.U digital persona fingerprinting. We develop a secure network system using a VPN on a specific domain. The confidentiality,

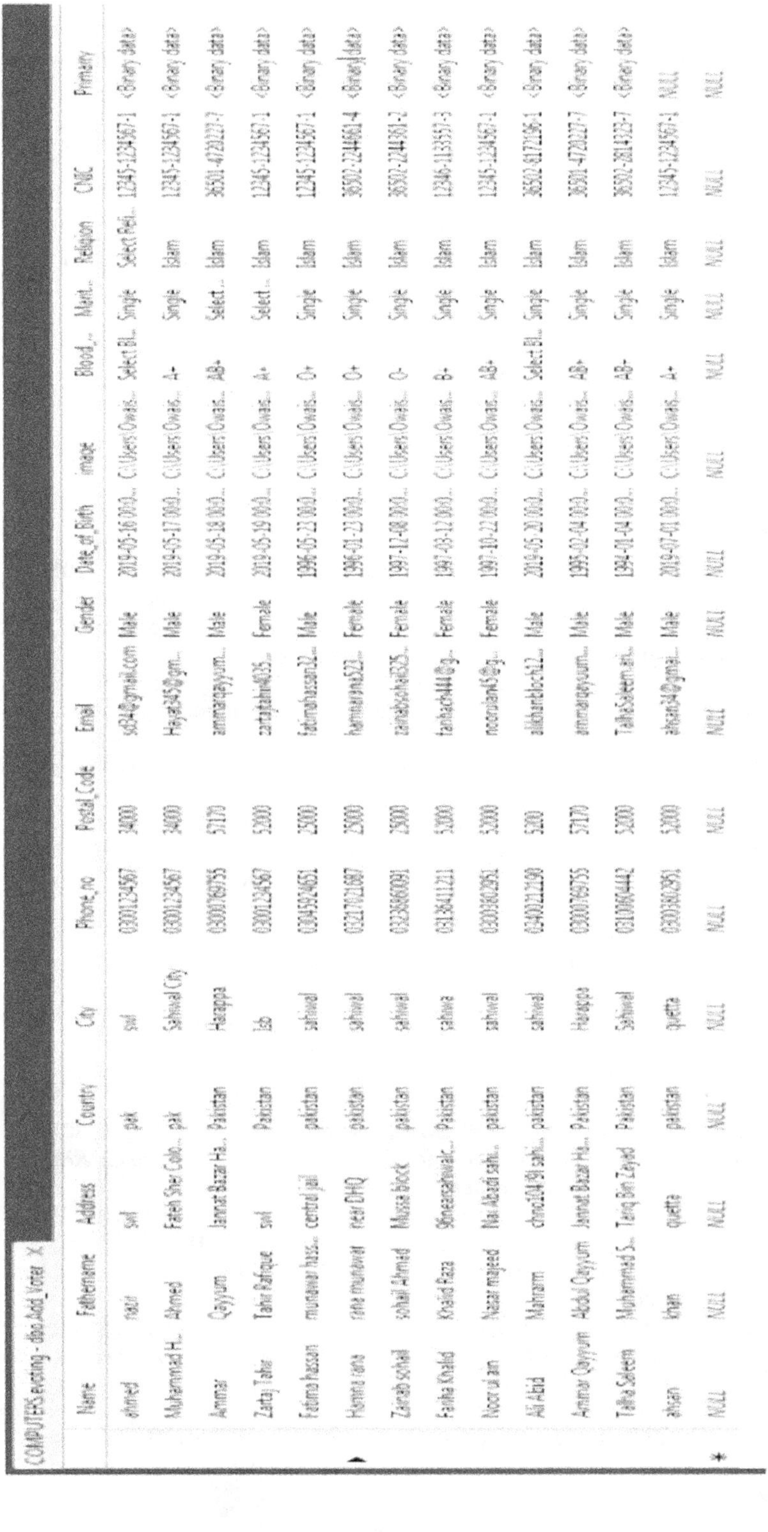

COMPUTERS evoting - dbo.Add_Voter X

Name	Fathername	Address	Country	City	Phone_no	Postal_Code	Email	Gender	Date_of_Birth	image	Blood_	Marit_	Religion	CNIC	Primary
ahmed	riaz	sol	pk	sol	03001234567	34000	sd54@gmail.com	Male	2019-05-16 00:0...	C:\Users\Owais...	Select BL..	Single	Select Reli..	12345-1234567-1	<Binary data>
Muhammad H...	Ahmed	Fateh Sher Colo...	pk	Sahiwal City	03001234567	34000	Hayat345@gm...	Male	2019-05-17 00:0...	C:\Users\Owais...	A+	Single	Islam	12345-1234567-1	<Binary data>
Ammar	Qayyum	Jannat Bazar Ha..	Pakistan	Harappa	03000769755	57170	ammarqayyum...	Male	2019-05-18 00:0...	C:\Users\Owais...	AB+	Select...	Islam	36501-4729227-7	<Binary data>
Zartaj Tahir	Tahir Rafique	sol	Pakistan	sb	03001234567	52000	zartajtahir4026..	Female	2019-05-19 00:0...	C:\Users\Owais...	A+	Select...	Islam	12345-1234567-1	<Binary data>
Fatima hassoon	murtasep hezz...	central jail	pakistan	sahiwal	03045924651	25000	fatimahussain22...	Male	1996-05-23 00:0...	C:\Users\Owais...	O+	Single	Islam	12345-1234567-1	<Binary data>
Hamna rana	rahe imyazawat	near DHQ	pakistan	sahiwal	03217020287	25000	hamnaryna322...	Female	1996-01-23 00:0...	C:\Users\Owais...	O+	Single	Islam	36501-2244661-4	<Binary data>
Zainab sohail	sohail Ahmed	Mussa block	pakistan	sahiwal	03284806091	25000	zainabjowhail325...	Female	1997-12-08 00:0...	C:\Users\Owais...	O-	Single	Islam	36501-2241581-2	<Binary data>
Fariha Khaled	Khalid Raza	S6heezeetwairak...	Pakistan	sahiwa	03136411211	53000	fazriha0444@g...	Female	1997-03-12 00:0...	C:\Users\Owais...	B-	Single	Islam	12345-11335173	<Binary data>
Noor ul ain	Nasar majeed	Nai Abddi sahh...	pakistan	sahiwal	03008600951	53000	noorsolain5@g...	Female	1997-10-22 00:0...	C:\Users\Owais...	AB+	Single	Islam	12345-1234567-1	<Binary data>
Ali Abid	Mahram	chnd184 9li sahh...	pakistan	sahiwal	03401212290	5200	aliicharblochl2...	Male	2014-05-30 00:0...	C:\Users\Owais...	Select BL..	Single	Islam	36501-81721896-1	<Binary data>
Ammar Qayyum	Abdul Qayyum	Jannat Bazar Ha..	Pakistan	Harappa	03000769755	57170	ammarqayyum...	Male	1995-02-04 00:0...	C:\Users\Owais...	AB+	Single	Islam	36501-4729227-7	<Binary data>
Talha Saleem	Muhammad S...	Tariq Bin Zayed	Pakistan	Sahiwal	03100664442	52000	TalhaSaleem an...	Male	1994-01-04 00:0...	C:\Users\Owais...	AB-	Single	Islam	36501-2814327-7	<Binary data>
ahsan	khan	quetta	pakistan	quetta	03003602951	52000	ahsan34@gma...	Male	2019-07-01 00:0...	C:\Users\Owais...	A+	Single	Islam	12345-1234567-1	NULL
NULL	NULL	NULL	NULL	NULL	NULL	NULL	NULL	NULL	NULL	NULL	NULL	NULL	NULL	NULL	NULL

FIGURE 12.18 Database view for hash value.

integrity, and authentication of e-voting is enhanced by using unique identification of fingerprints of the voters through the conversion of an image to binary data templates. After the election process, effectual results are displayed via an auto-count process that is less time consuming, cost effective, and increases the voting rate. System also includes the messaging service for voter satisfaction to inform the voter that their votes have been counted and cast successfully. This system reduces the tempering rate and enhances the security and transparency of the election process.

12.7 FUTURE WORK

As a future line of work, we will enhance the different security checks and accuracy features of the system by introducing the cloud computing in e-voting system using blockchain.

REFERENCE

[1] Brereton, R.C., Why democracy is important. In *Global Ethics Network*, 2018.
[2] Musleh, M.H., What is the concept of democracy? In *ResearchGate*, 2019.
[3] Chaudhry, A., U. Mazher, and M.H. Khan, How socio-economic conditions affect voting turnouts in pakistan? *A District-Level Analysis,* 2018.
[4] Khan, N.U., and S. Akhtar, Historical challenges to Pakistan's good governance: Reforms in the election process in conducting free, fair and transparent elections abstract. *JPUHS*, 29(2), July–December 2016.
[5] Group, I.C., *Reforming Pakistan's Electoral System Executive Summary and Recommendations,* p. I, 2011.
[6] Kelsey, J., Regenscheid, A., Moran, T., & Chaum, D. Attacking paper-based E2E voting systems. In *Towards Trustworthy Elections*, Springer, pp. 370–387, 2010.
[7] Dev Ananda, A.B., Janet Gonzalez, and Martha Prempeh, *The Future of E-Voting.* Information Technology and Public Policy, 2019.
[8] Olaniyi, O.M., Bala, J. A., Ndunagu, J., Abubakar, A., & Is' Haq, A. *V-Authenticate: Voice Authentication System for Electorates Living with Disability,* 2019.
[9] Nakirya Brenda Kintu, M.I.Z., A secure E-voting system using biometric finger-print and crypt-watermark methodology. In ASCENT International Conference Proceeding-International Systems and Engineering, 2018.
[10] Shaheen, S., M. Yousaf, and M. Jalil, *A Smart Card Oriented Secure Electronic Voting Machine Built on NTRU,* 2018.
[11] Sreerag, M., R. Subash, Vishnu C. Babu, Sonia Mathew, and Reni K. Cherian, Aadhaar based voting system using android application. *International Research Journal of Engineering and Technology (IRJET)*, 05(04), 2018.
[12] Hjálmarsson, F.Þ., et al., Blockchain-based e-voting system. In 2018 IEEE 11th International Conference on Cloud Computing (CLOUD), 2018. IEEE.
[13] Alsaade, F, Effectiveness of score normalisation in multimodal biometric fusion. *Journal of Information & Communication Technology (JICT)*, 3(1), 2009.
[14] Sundar, N. A., P.M.V. Kishore, and Ch. Suresh, A secure E-Voting system using RSA and Md5 algorithms using random number generators. *International Journal of Applied Engineering Research*, 13, 9468–9473, 2018.
[15] Darwish, A., and M. Gendy, A new cryptographic voting verifiable scheme for E-voting system based on bit commitment and blind signature. *International Journal of Swarm Intelligence and Evolutionary Computation*, 6(2), 2017.

[16] Dongre, R., S. Pal, S. Dhurve, N. Surywansi, and K. Itankar, A review paper on fingerprint based voting machine. *International Journal of Advanced Research in Electrical, Electronics and Instrumentation Engineering (IJAREEIE)*, 6(4), 2017.

[17] Kharmate, Balkrushna Bhagwatrao, Shahebaz Shakil Shaikh, Prashant Ravindra Kangane, and Tushar Anant. A survey on smart E-voting system based on fingerprint recognition. *International Journal of Innovative Research in Computer and Communication Engineering*, 3(9), 2015.

[18] Tornos, J.L., J.L. Salazar, and J.J. Piles, An eVoting platform for QoE evaluation. In *2013 IFIP/IEEE International Symposium on Integrated Network Management (IM 2013)*, 2013. IEEE.

[19] Olaniyi, O.M., et al., Design of secure electronic voting system using fingerprint biometrics and crypto-watermarking approach. *International Journal of Information Engineering and Electronic Business*, 8(5), 9, 2016.

[20] Khan, Adnan Alam, and Abdul Aziz, Face recognition techniques (FRT) based on face ratio under controlled conditions. In *2008 International Symposium on Biometrics and Security Technologies*, pp. 1–6, 2008. IEEE.

13 Blockchain-Based Supply Chain System Using Intelligent Chatbot with IoT-RFID

Khurram Shahzad, Hasnat Ahmed, Faraz Ahsan, Khalid Hussain, and M N Talib

CONTENTS

13.1   INTRODUCTION

The blockchain is an emerging technology that is transforming our industries rapidly. The blockchain is the technology behind Bitcoin, the distributed-ledger system that records transactions. Every block contains the ledger, and all blocks are onnected with one another in a chain with the support of cryptography. The machines (nodes) are connected with one another, and each node contains a copy of the ledger. Each block is secured because every new block will contain the hash of the previous block, making tampering practically impossible. The blockchain can provide us transparency and traceability, along with security. The blockchain also acts as a distributed database. What is supply chain? Basically, the supply is how the products will travel from supplier to manufacturer and then to the end consumer. The process involves many steps that involve how the products will travel from one phase to another. But the modern supply chain is very complex due to transparency and traceability issues. The modern supply chain includes the paperwork at each step needed to process each and every step.

The artificial intelligence means how can we make the machine intelligent, how machines will mimic humans. In simple terms, the artificial intelligence is the simulation of human intelligence. The process involves training, predicting, and many other things to make the machine smarter and intelligent. So there are many examples available, like chatbot, problem solving, speech recognition, predicting, and training, etc. This study is conducted to focus on how chat-bot will act more secure and reliable than a normal one, having a combination of machine learning and natural language processing.

IOT, the internet of things, means that different devices are connected to one another over the internet. The devices/objects also have the ability to exchange the information and act upon it. There are different sensors, chips, and other protocols embedded in IOT to receive and send information. We will focus on what is RFID, the radio-frequency identification. The RFID is a chip, or digital tag, that can be attached to any product. The tags contain the digitally stored information of the product that is further stored on blockchain. It uses the electromagnetic fields to automatically identify and tracks the products by no means.

The blockchain, along with artificial intelligence and IOT, has the ability to become the universal supply chain operating system, providing security, transparency, and scalability. The technology blockchain allows us to secure and have transparency with all types of transactions. Each time the product changes locations or owner, the transactions could be written on the ledger in the block. It maintains the overall history of the product from where the product is dispatched and who receives it. Each and every thing is written on the blockchain, which is cryptographically secured. This could reduce the time delays, human efforts, and paperwork every time.

13.2 BACKGROUND

The blockchain is an emerging and more disruptive technology. The brain behind this technology is a person or group of persons named by Satoshi Nakamoto [1]. In 2008, the Satoshi Nakamoto published a paper titled "a peer-to-peer Electronic cash system." From that day, it evolved into something bigger day by day, and the main question that everyone is asking is, what is blockchain? The blockchain technology created a new type of internet. The first financial use of blockchain is Bitcoin, and now other tech companies are trying to find out other potential uses for this technology. The Bitcoin network was created on January 3, 2009, when Nakamoto mined the first block of chain, which is also known as the genesis block. Bitcoin is a digital currency that uses peer-to-peer technology. It is the first decentralized payment network that is powered by users (miners). It doesn't involve any central authority or banks to manage this currency. Bitcoin is open-source and publicly available; nobody owns it, and everyone can participate in it this network.

A blockchain is a distributed ledger that guarantees immutability and trustworthy transactions that can't be changed (any value assets). It provides the secure and reliable environment through a consensus protocol that is secured by cryptography. It is immutable, transparent, and open source that anyone can voluntarily participate in and leave as he will. There are two broad types of blockchain: public (anyone can participate) and private (required permission before participating in the network). If we compare it to centralized systems, the blockchain will remove cost, saves time, reduce risk, and increase trust.

The blockchain is also a distributed database. It is a way of using a blockchain network that also has obvious benefits. The data on blockchain isn't stored on a single location, meaning that it shared the data across all the nodes on the network, which is cryptographically secured and easily verifiable. The data on the network is hosted by millions of computers; everyone can see the transaction details, but no one can alter them because it is temper proof. Each block on chain is secured because every new block will contain the hash of the previous block, making tampering practically impossible.

In public blockchain, everyone can participate in the network, and anyone can leave it. The information on public blockchain is shared among all the parties. If all nodes agreed, transactions will then be added on blockchain. If someone tries to alter the block, the other nodes will deny it. It uses a proof-of-work algorithm to find the hash. The miners solve the complex mathematical puzzle to add the new blockchain in the chain. After that, they get the rewards in the form of digital currency. We will discuss Ethereum, how this platform is using blockchain and how can we build a dApp on these.

The private blockchain is a permissioned distributed network. If someone wants to join the network, first he must request for permission. The information stays within the network; it can't be shared with anyone else who is not in the network. Private blockchains are also cryptographically secured and temper proof. This is more useful for enterprises to use private blockchain. We will discuss hyper ledger fabric.

13.2.1 ETHEREUM

Ethereum is a decentralized platform that is based on blockchain technology; it allows the developers to build and deploy their decentralized applications [2]. Ethereum is an open source platform. Ethereum is not similar to Bitcoin, but Bitcoin and Ethereum are both distributed-ledger public blockchain networks. Ethereum platform runs smart contracts, which means the application will run the same as it is programmed without any third-party interference, censorship, or possibility of downtime. This allows the developers to create applications like markets, storing registries and moving funds from one point to another, and many other things that are not yet invented, all this without any third party or middleman involvement or risk.

In Ethereum blockchain, the miners work to earn ether, the new type of crypto token that is used to fuel the blockchain network instead of only mining for Bitcoin. The developers use ether to pay for transactions fees and other services on the Ethereum network. There is also a second type of token on Ethereum blockchain that is used to pay miners' fees for adding transactions in the block; it is called gas. On every execution of a smart contract, there is certain amount of gas required to be sent along.

The core innovation of Ethereum blockchain is its Ethereum virtual machine (EVM), which is a Turing software that runs on the Ethereum network. The Ethereum virtual machines simplify the process, making the blockchain application easier and more efficient. Instead of building an entire new blockchain from scratch for each new application, Ethereum allows the development of thousands of different applications to be run on one platform, which is powered by public blockchain Ethereum.

A. **Smart Contract**

In simple words, the smart contract refers to as the agreement or contract between two parties. But, technically, it is a digital contract based on a computer program that can help us to exchange assets or anything that is valuable. On blockchain, blockchain, the smart contract in like an operating computer program, which executes automatically when the same conditions are met. As we We know that the smart contract will run the same as it was programmedprogrammed, without any middleman interference, fraud, and downtime.downtime [3].

So what makes the Ethereum different? Every blockchain has the ability to process the code, but on Ethereum, the developers can make whatever they want. This means developers can build as many applications as they want, which can be beyond anything that we ever have seen before. The Ethereum blockchain allows the programmers to build and deploy their decentralized application. It can be also be used to build decentralized autonomous organizations (DAO). A DAO is fully decentralized and autonomous with no single leader.

B. **Solidity**

The smart contract is in written in a specific language, which is called solidity. The solidity is a high-level programming language for implementing and writing smart contracts. The goal is to target the Ethereum virtual machine. It was influenced by Python, C++ and JavaScript. Solidity is a case sensitive, statically typed programming language which supports inheritance, libraries, and complex user-defined types.

13.2.2 HYPERLEDGER FABRIC

The Hyperledger is developed under Linux foundation, the open-source distributed ledger framework. The main goal is to create an enterprise-level application to advance cross-industry blockchain technologies. And Fabric is the first distributed-ledger platform that has the ability to write smart contracts in general languages, such as node, java, go and JavaScript, rather than specific languages. Our primary focus is on the consensus about how the information will be validated.

Hyperledger Fabric is generic. It is developed under Linux foundation and now IBM is managing it. It provides long-term development support. As compared to other blockchain networks like Bitcoin and Ethereum, the Hyperledger fabric is more suitable for a broad range of applications. Hyperledger Fabric is a purely private blockchain. The smart contract systems are preferred because they don't used crypto tokens. However, they can create the crypto token from this. It provides high throughput performance (can do a half million transactions per min.). The data will be immutable, and every node in the network will have the copy of ledger. The ledger contains the transaction and other data as well. The information can't be modified or deleted.

13.3 CHATBOT WITH BLOCKCHAIN

Chatbot and blockchain are made for each other, and they are a perfect match. The blockchain technology is our first choice when we talk about trust. The combination of chatbot and blockchain will automate the marketing activity. It will help us communicate with our clients via a trusted chatbot that is totally based on blockchain technology [4–6].

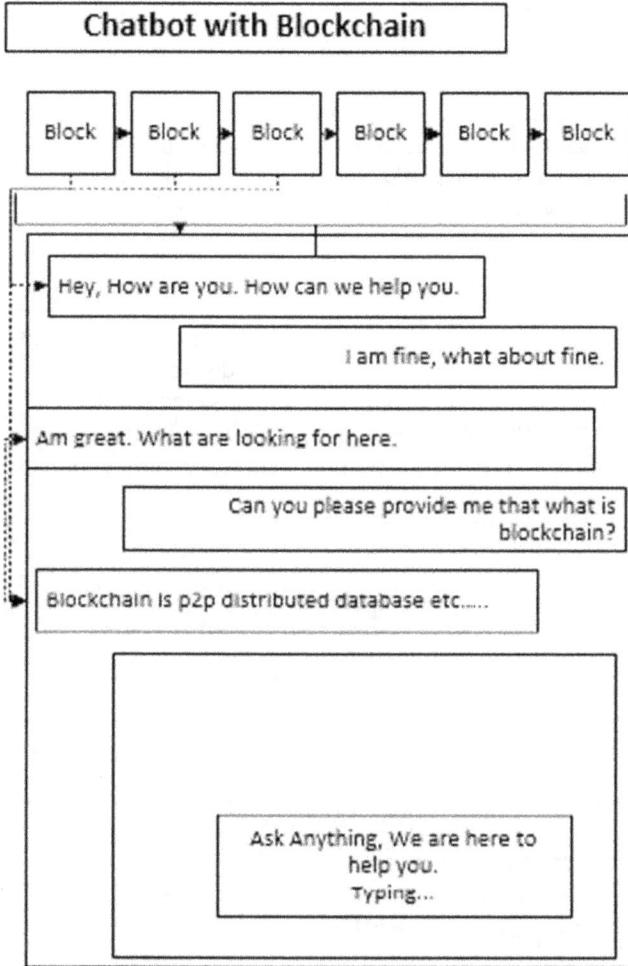

13.3.1 IOT (INTERNET OF THINGS)

In simple words, the "internet of things" or IOT, is defined as the network of internet-connected devices that have the ability to collect and exchange information. There are two parts in internet of things: the internet is the backbone of connectivity, and "things" means any object or device. To simplify it more, the IOT means taking all the objects/devices in the world and connecting them together via internet. The ability to send or receive data makes the objects smarter. The IOT is an architectural framework that allows integration, and data will be exchanged between the physical world and computer systems over network infrastructure. In IOT, all the devices that are connected to the internet can be categorized as follows:

- The things that collect data and then send it.
- The things that receive data and act upon it.
- The things that can do both.

13.3.2   RFID Chips

The RFID (radio-frequency identification) is the form of wireless communication that uses electromagnetic fields to automatically detect and track the tags on the product. The tags are used in many fields. For example, an RFID tag is attached to a product in supply chain to track the location and to verify the digitally stored information in the tags [7].

13.3.3   RFID with Blockchain

The combination of RFID and blockchain provides many advantages in supply chain. They can be used for identification of products, inventory control, distribution, automatic tracking, and testing of the supply chain. The information will be stored on blockchain as from source to destination. The blockchain also makes it possible to check the information for specific things such as where it is from and what is the current location.

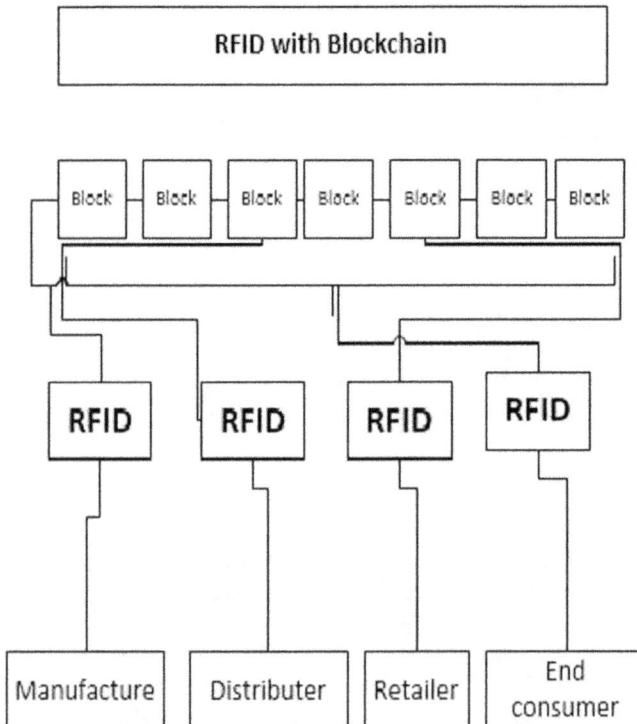

13.3.4 MODERN SUPPLY CHAIN

In simple terms, the supply chain is the whole process of products from producer to manufacturer and manufacturer to end consumer. Today's supply chain is very complex. We are doing a lot of paperwork. Each and every member is maintaining his own records. From the manufacturer to the receiver, records are complex and everyone does paperwork to verify success. At the end, we actually don't know whether the product is original or not. We even can't identify the product's origination. There is a lot of paperwork in modern supply chains to ensure each and every thing [8].

Paperwork captures where products are dispatched and where they have been all the time [9]. If a product encounters problems, we can't even track it back easily because it involves complex paperwork to check each and every product. In the supply chain, we must be careful from the manufacturer who is receiving the raw material and the receiver who is buying the actual product. Each and every thing should be secured. For example, in custom work, we must do a lot of paperwork to get our product back. If we are sending some goods to another country, shipping requires that we pass the paperwork to verify each and every thing. The process is very time consuming [9].

13.4 PROBLEM STATEMENT

Today's supply chain is a very long and complex process that involves many steps from producer to end consumer. There are many problems that we face while ensuring shipments and storing the records of the products. If someone lost the product and wants to track it down, it takes a very complex system. We face problems to trace and track the products because we don't have a proper tracking system. A lot of paperwork must be done in the supply chain. We don't know actually where the product comes from and where it has been all the time. Each and every organization has its own records, which don't contain the actual information. The main issue is the traceability and transparency issues in the supply chain.

For example, recently I lost a bag while traveling to Pakistan from Greece. When I got back to the airport security to complain about it, they simply asked me to fill out the papers and submit to them. Even we don't have proper tracking systems at airports. Even we can't find the bag; they said that they don't know about it [10].

13.5 LITERATURE REVIEW

Saveen A. Abeyratne, Radmehr P. Monfared (UK 2016) blockchain ready manufacturing supply chain using distributed ledger. The purpose of this paper to provide the overall concept about how to use the supply chain with blockchain. They have described that every user must register on the applications that will be further verified by the registrar to complete the registration process and will gain the access to the website. They have explained the whole process about how the supply chain will be worked from the manufacturer to the supplier. They have taken the example cardboard boxes to explain how can we implement this with blockchain. The whole process involves the manufacture to retailers.

Simon Figorilli, Francesca Antonucci (2018) write about a blockchain implementation prototype for the electronic open source traceability of wood along the whole supply chain. The work defines the implementation of blockchain architecture with electronic traceability technology, or RFID. The RFID open source technology will make the key difference. It defines the whole process from the timber making to final consumer. The RFID will be useful for the traceability of products, and blockchain can be used to store information that is unchangeable and secured. The blockchain can also help to achieve the security requirements with the help of IOT. Different kinds of IOT tags were used, and the main aim is to collect the information about the products and store it in the blockchain.

Kari korpela, Jukka Hallikas (2017) (Digital supply chain transformation toward blockchain integration) The paper defines the transformation of the digital supply chain with integration of blockchain technology. The main aim was to establish and get an overview about how B2B digital supply chain can be accelerated. Many experienced business managers from different companies were asked to create the requirements and functionalities that help in business process integration. The main ideas discuss how the blockchain technology can help to accelerate the business process integration in the digital supply chain. In current supply chain, many requirements and functionalities are missing that can be solved by embedding the blockchain technology in the digital supply chain.

Aaron Benningfield (2017) [11,12] studied the Hyperledger–supply chain traceability and anti-counterfeiting, exploring how we can use Hyperledger to ensure the transparency and accuracy about the products, customers, regulators, and partners, etc. The different cases were described in which the duplication and traceability were addressed. They used the distributed ledger technology (DLT) Hyperledger to provide quick updates and tracing to proceed accordingly. It had the ability to track the product from start to end-point delivery while maintaining the whole information and control to other nodes as well. The three main components are used ensure the end-to-end traceability, regulatory concern, and physical characteristics.

Kristoffer Francisco and David Swanson (5 January 2018) Technology Adoption of Blockchain for Supply Chain Transparency. In this paper, the authors used blockchain technology to improve the transparency of the supply chain and to address some common issues like counterfeiting, negativity, corruption. They suggest the use of the technology of unified theory of acceptance and use of technology (UTAUT) and the idea of innovation adoption as a base framework to overcome all these issues for the supply chain. If you want to improve it further, describe how blockchain adopts the type of product and service impact on the end user. How can IoT, the internet of things, provide information and integrate the blockchain?

13.6 METHODOLOGY

Now, what can we do with this? How can we resolve this problem? What are certain ways to reduce the paperwork and time? We came up an idea to use blockchain technology in the supply chain with the help of chat bots and RFID chips that help us to solve this problem easily and efficiently. Each and every product is attached with a tag that contains its information. The tag will be in the form of a barcode, RFID, or QR code. This tag will represent the unique digital cryptographic identity that links the product to its virtual identity on the blockchain network.

Blockchain has the ability to become the universal supply chain operating system providing security, transparency, and scalability. The technology blockchain allows us to secure and ensure the transparency of all types of transactions. Each time products change locations or owners, the transactions could be written on the ledger in the block. So it maintains the overall history of the product from where the product is dispatched and to whom receives it. Each and every thing is written on the blockchain that is cryptographically secured. This could reduce the time delays, human efforts, and paperwork every time [13].

Every user has their own digital profile on the network, which will be created upon registration. The profile displays their information, such as their location, description, certification, and association with their products. Every product that is signed by the user would also have a link from the product profile to the user's profile. Our system allows the users to change their privacy as well. The users can choose to remain anonymous from everyone, but each must be certified by a registered auditor or certifier to maintain the trust of the system.

Structure:

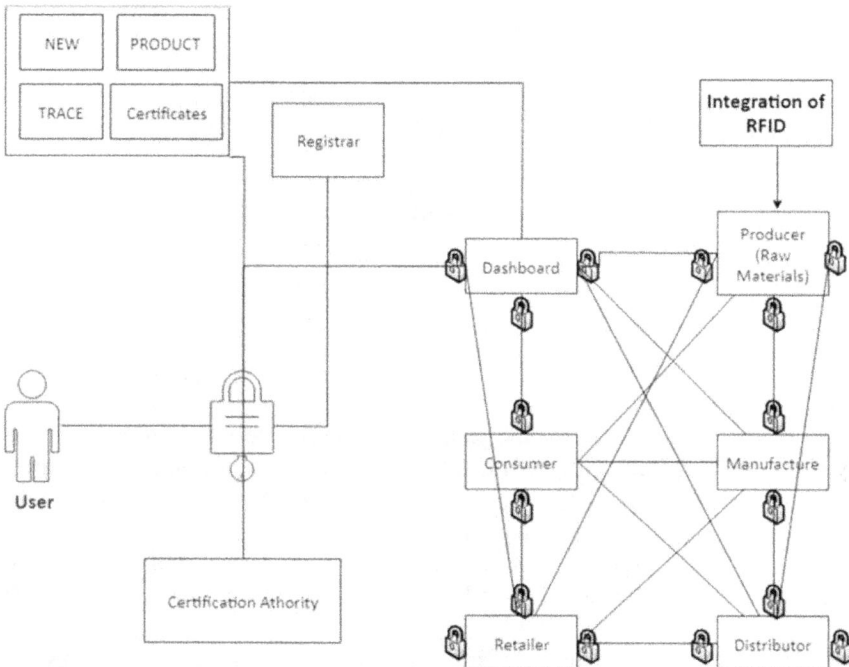

Architecture:

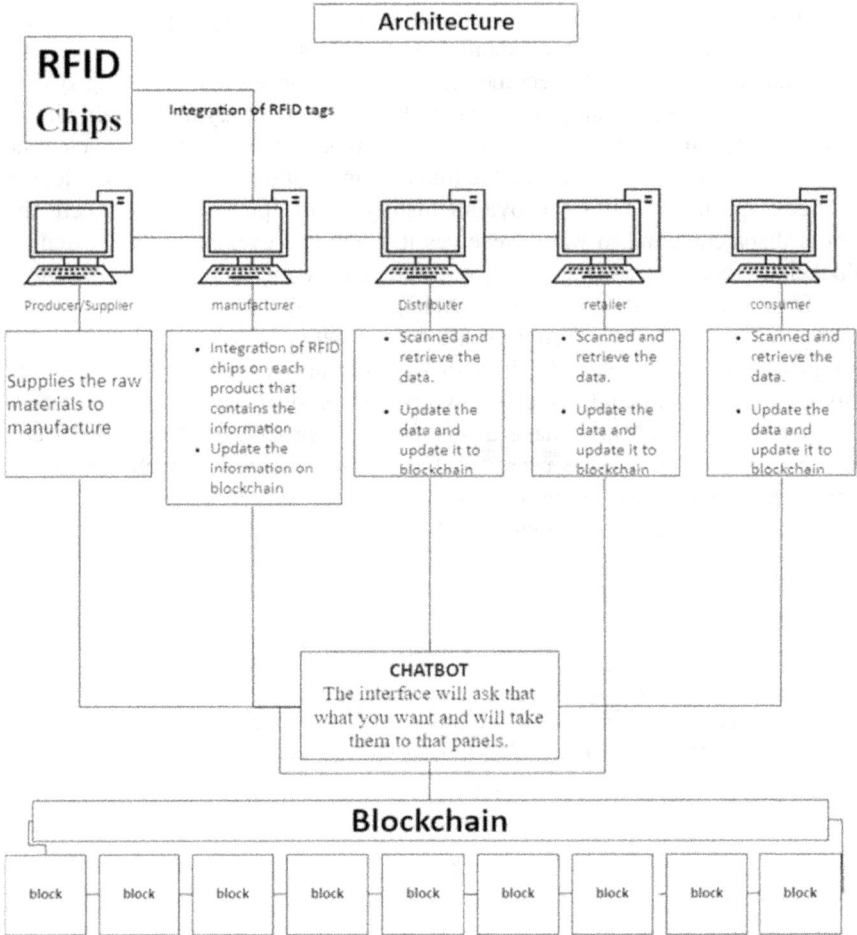

To explained the potential for the proposed concept, the whole process of the supply-chain-ready system with the integration of blockchain, artificial intelligence, and IOT is considered for the supplier (raw materials from where comes from), manufacture, distributor, retailer, and end consumer.

In this application, large numbers of users are involved. The product will be in any form, and we will discuss the overall scenario about how the system will work from producer to end consumer.

Everyone mentioned in this scenario must have to register themselves on the system through registrar and should have their own unique identity on the blockchain network. Each user will interact with the graphical user interface of the system that will be represented as chatbot from where you can easily ask what you want to do and the system will come up with that interface.

1. Producer: The raw materials come from producer to the manufacturer. When the raw materials are dispatched, the origin is noted, other information is reported as well, and further information is stored on blockchain. The user opens the application and the chatbot interface. The chatbot will ask who they are and what they want to do on the application. As everyone has a unique identity on the system, the producer will also access the application to enter the records. He will update the location, name, date, and other things related to the product, and the manufacturer should know where the product has comes from.

2. Manufacturer: The producer sends the raw materials to manufacturer. The manufacturer opens the system with the chatbot interface, and then he will request to update the status that the manufacturer receives the goods and updates the status to say that the material has been received. Now the manufacturer will process the raw materials and convert it into some form of product. The integration of RFID will be done by manufacturer as well. The RFID tags contain the digital information that is further stored on the blockchain. Again, the status will be updated about the products through their profile. Now the distributer will take the products to dispatch to various retailers. The status will be changed, noting that the products have been shipped to distributor.

3. Distributer: The next one is the distributer, who transports the products on various locations via shippers or containers. The distributer will receive the products and update the status on the network, noting that the products have been received. This will make it easier for the products to be updated quickly and efficiently. The blockchain will be automatically updated with the information which enables the traceability of products in larger batches. Now the status will again change, noting that products have been shipped to retailers.

4. Retailers: The retailers will receive the full packed boxes of product, remove the boxes, and update the statues with the current status. Now the retailers will send the products to the consumer through locations where the customer can buy it. The retailer will simply scan the packages to update the details to where it has been dispatched. Now he will update the status that the products have been transferred to consumer.

5. Consumer: The consumer will received the products from retailers and update the status that products have been received. Now consumers will scan the RFID tag to retrieve the whole information of the product, from the producer to consumer.

Product Image

Batch Number

Product ID

Timestamp (Date and Time)

Certificate generated with each product

General details

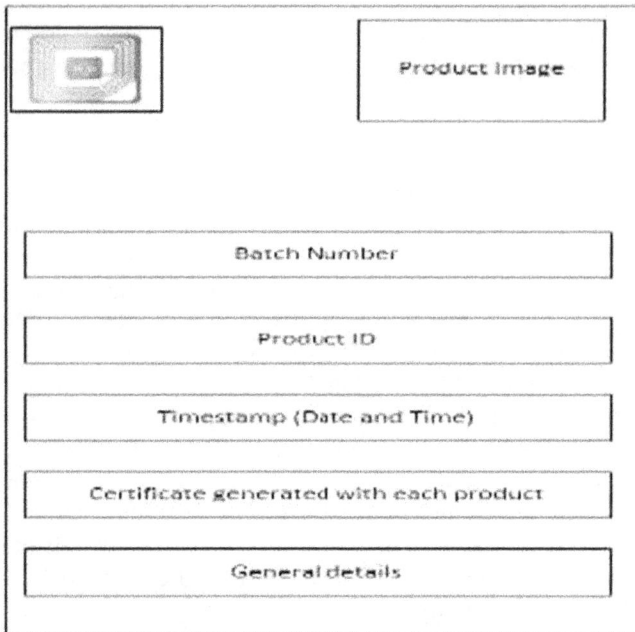

13.7 CONCLUSION

The blockchain ready supply chain with an integration of artificial intelligence and IOT has changed the whole scenario for this system. The whole new system that provides the security, transparency, scalability, and much more. Each and every thing is secure, and there is no need to be worry about the loss of data. Everyone in the network will be maintaining a copy of the ledger, which is also cryptographically secured and temper proof. This makes the system efficient and returns better results as compared to other systems. The blockchain reduces the paperwork and intermediaries, which helps to make a trust-less environment system. The blockchain can solve many problems with the help of artificial intelligence and IOT to ensure the security of the system [14,15].

Now many industries are involved to move on these technologies and build an appropriate solution. These technologies are transforming the industries rapidly in many sectors, such as financial, digital identity, smart property, and other potential application.

REFERENCES

[1] Singh, Akhilendra Pratap, Nihar Ranjan Pradhan, Shivanshu Agnihotri, N. Z. Jhanjhi, Sahil Verma, Uttam Ghosh, and D. S. Roy. (2020). "A novel patient-centric architectural framework for blockchain-enabled healthcare applications." *IEEE Transactions on Industrial Informatics,* 17(8).

[2] Buterin, V. (2014). "Ethereum white paper: A next generation smart contract & decentralized application platform." [Online]. Available: http://blockchainlab.com/pdf/Ethereum_white_paper.

[3] Humayun, M., N. Z. Jhanjhi, B. Hamid, and G. Ahmed. (2020). "Emerging smart logistics and transportation using IoT and blockchain." *IEEE Internet of Things Magazine*, 3(2), 58–62.

[4] Chatbots, B. W. L. (2017). "Chatbot magazine." [Online]. Available: https://chatbotsmagazine.com/chatbots.

[5] Ali, Sadia, Yaser Hafeez, Mamoona Humayun, Nor Shahida Mohd Jamail, Muhammad Aqib, and Asif Nawaz. (2021). "Enabling recommendation system architecture in virtualized environment for e-learning." *Egyptian Informatics Journal.*

[6] Nath, J. (2018). "Chatbots magazine." [Online]. Available: https://chatbotsmagazine.com.

[7] R. Journals. (2018). "RFID and blockchain are changing supply chains." *Hatkinson-Kent, Luke,* 1–2.

[8] Latif, Rana M. Amir, Khalid Hussain, N. Z. Jhanjhi, Anand Nayyar, and Osama Rizwan. "A remix IDE: Smart contract-based framework for the healthcare sector by using Blockchain technology." *Multimedia Tools and Applications,* 1–24.

[9] Gajendran, N. (2020). Blockchain-based secure framework for elearning during COVID-19. *Indian Journal of Science and Technology*, 13(12), 1328–1341.

[10] Brody, P. (2017). "EY building a better working world." *Digitalist Magazine by SAP.*

[11] Tillström, T., and Erik Hillbom. (2016). "Applications of smart-contracts and smart-property utilizing blockchains." [Online].

[12] A. B. (2017). "Hyperledger - supply chain traceability: Anti counterfeiting." [Online].

[13] Kulkarni, A. (2018). "About us: Indorse." [Online]. Available: https://blog.indorse.io/@avadhoot_kulk.

[14] Marr, B. (2018). "Forbes." Forbes Media LLC, [Online]. Available: https://www.forbes.com.

[15] Alamri, Malak, N. Z. Jhanjhi, and Mamoona Humayun. (2019). "Blockchain for Internet of Things (IoT) research issues challenges & future directions: A review." *International Journal of Computer Science and Network Security,* 19, 244–258.

Index

For Product Safety Concerns and Information please contact our EU
representative GPSR@taylorandfrancis.com
Taylor & Francis Verlag GmbH, Kaufingerstraße 24, 80331 München, Germany

www.ingramcontent.com/pod-product-compliance
Lightning Source LLC
Chambersburg PA
CBHW060350220326
41598CB00023B/2872

9 7 8 1 0 3 2 2 0 3 7 5 1